Physics

Physics: An Introduction to Physical Dynamics provides an accessible introduction to the fundamentals of physics for science and engineering undergraduates who are studying elementary physics.

This textbook contains 12 chapters with accompanying problem sets and explains the dynamical properties of a variety of physical systems. The first six chapters introduce Newton's laws of motion, followed by the concepts of mechanical work and mechanical energy, with illustrative applications to the translational and/or rotational motion of inflexible objects such as particles and 3D objects of fixed shape. The next four chapters generalize the application of Newton's laws and mechanical energy to flexible systems, including flowing fluids, waves on strings, and oscillating springs. The last two chapters elucidate the laws of thermodynamics, especially heat energy transfer between systems at different temperatures.

Some familiarity with topics in elementary mathematics, including calculus, is assumed. A wide variety of situations are explored, by means of which a student should acquire an enhanced understanding of the properties of physical systems from the astronomic scale to the microscopic.

Key Features

- Covers the classical mechanics of both single particles and assemblies of particles subject to forces.
- Contains wide-ranging sets of examples and worked problems.
- Covers much of the material that a student might expect to encounter during the first year of a university physics course.

Physics

An Introduction to Physical Dynamics

Michael J.R. Hoch

CRC Press
Taylor & Francis Group
Boca Raton   London   New York

CRC Press is an imprint of the
Taylor & Francis Group, an **informa** business

Designed cover image: © Shutterstock

First edition published 2025
by CRC Press
2385 NW Executive Center Drive, Suite 320, Boca Raton FL 33431

and by CRC Press
4 Park Square, Milton Park, Abingdon, Oxon, OX14 4RN

CRC Press is an imprint of Taylor & Francis Group, LLC

© 2025 Michael Hoch

ISBN: 9781032779478 (hbk)
ISBN: 9781032767505 (pbk)
ISBN: 9781003485537 (ebk)

DOI: 10.1201/9781003485537

Typeset in Times
by Newgen Publishing UK

Contents

Preface

Classical mechanics is an important branch of physics and finds many applications in both science and engineering. This book is aimed at beginning physics students at universities. The students will have completed, or will be attending, introductory mathematics courses.

Historically, great interest in the dynamics of moving objects followed Isaac Newton's major contributions to the subject in the seventeenth century. Using his inverse square law of gravitation, Newton was able to successfully account for Kepler's laws of planetary motion around the Sun. This achievement inspired many other applications of what were called Newton's laws of motion.

This book, which contains twelve chapters with accompanying problem sets, deals with the dynamical properties of a variety of physical systems.

The first three chapters introduce the basic ideas of mechanics – in particular mass, length, and time, with Chapter 3 exploring the concept of momentum as it relates to Newtonian mechanics.

The application of Newtonian mechanics is addressed in Chapters 4–6. Chapter 4 introduces the concepts of mechanical work and mechanical energy, with illustrative applications to the translational and/or rotational motion of inflexible objects such as particles and 3D objects of fixed shape. Chapter 5 discusses the important special case of rotational motion, while Chapter 6 covers rigid body dynamics.

Chapters 7–10 generalize the application of Newton's laws and mechanical energy to flexible systems, including flowing fluids, waves on strings, and oscillating springs.

The final two chapters elucidate the laws of thermodynamics, especially heat energy transfer between systems at different temperatures.

The Système International (SI) units for length and time are introduced in Chapter 1, together with the recent fundamental definition of the mass unit. SI units are used throughout the book.

Acknowledgment

To my family, for their continuing support in preparing the book for publication.

I particularly wish to acknowledge my wife Renée, for her support over the entire duration of the project, and my eldest son Andrew and his wife Lucy, for their assistance.

About the Author

Michael J.R. Hoch spent many years as a visiting scientist at the National High Magnetic Field Laboratory at Florida State University, USA. Prior to this position, he was a professor of physics and the Director of the Condensed Matter Physics Research Unit at the University of the Witwatersrand, Johannesburg, where he is currently Professor Emeritus in the School of Physics.

1 The Physical World

1.1 INTRODUCTION

Physics holds major importance in our scientific endeavours to gain a deeper understanding of the world in which we live. Many discoveries made by physicists, particularly in the past few centuries, have, over time, led to technological innovations that have transformed our present-day lives. The boundaries of physics are not rigid, and many important interdisciplinary activities have emerged involving collaborations between physicists and other scientists in various fields. Mathematics is essential for developing the theoretical insights needed to interpret experimental observations. The range of physics activities is vast involving phenomena on length scales from sub-atomic to astronomically large. While this enormous range may appear daunting to an individual starting out in physics, the situation is helped by many unifying concepts and the establishment of fundamental laws of nature that underlie all physical phenomena in the universe.

This book engages in a presentation of classical mechanics and related subjects. A central topic involves the motion of objects subject to applied forces. As shown in later chapters, Newton's laws of motion are of fundamental importance in dealing with this dynamical behaviour of such objects. The introduction of the concepts of momentum and energy facilitate the discussion. In later chapters, the approach is generalized to rigid body motion, fluid properties, oscillations, and waves. The present chapter introduces the fundamental physical properties of mass, length, and time, together with the units used in the measurement of these properties. The last part of this chapter briefly reviews what is known about the fundamental forces of nature and the length scales over which they operate.

The basic concepts of mass, length, and time are familiar to us all from everyday experience. Over time, the internationally accepted definitions of the units for measuring these quantities have evolved and become increasingly precise particularly since the mid-twentieth century. Historically, the meter (m), kilogram (kg), second (s) system of units, abbreviated MKS, was established in France in the late eighteenth century and later adopted by many countries, particularly for scientific purposes. The standard kilogram was defined in terms of the mass of a cylinder of platinum–iridium

DOI: 10.1201/9781003485537-1

alloy kept in a safe in a vault in Paris. Copies of this standard were made available to other countries. Iridium in the alloy hardened the surface of the cylinder and reduced possible surface wear. For length, the meter was defined as the distance between two marks on a platinum–iridium bar kept at a particular temperature in a Parisian vault. This length was based on the distance from the equator to the North Pole and was chosen as one ten-millionth of this distance. It later turned out that the measured quadrant distance was somewhat in error, but the marked distance on the bar was retained as the definition of the meter. The second was based on the length of a solar day taken as 24 hours corresponding to 86,400 seconds. Actually, the length of the solar day varies during the year because of the slightly elliptical nature of the Earth's orbit around the Sun and the axial tilt, or obliquity, of the Earth's axis of rotation with respect to the ecliptic, which is the plane of the Earth's orbital motion. It is clear that the standards used in the original MKS system are not satisfactory because, firstly, they are based on macroscopic objects that are subject to possible change over time as the result of wear and, secondly, they can be measured with only limited precision. Starting in the mid-twentieth century, the basis for defining the MKS units has been altered from the macroscopic to the atomic scale where properties are stable over extremely long times. A fundamental physical constant, the speed of light in vacuum, is used, together with the new unit of time, in defining the new unit of length as described below.

The revised system as adopted by international agreement is known as the Système International d'unités, abbreviated as the SI system. In addition to the speed of light in vacuum, c, there are other fundamental physical constants such as the charges and masses of the electron, $-e$ and m_e, the proton, e and m_p, and the neutron, 0 and m_n, plus a number of other fundamental quantities, including Planck's constant h, Avogadro's number N_A, and Boltzmann's constant k_B. A table of values of the fundamental constants is given in Appendix 1 and the values indicate the extremely high precision that has been achieved in their determination. These constants play a crucial role in comparing theoretical predictions with experimental findings. It follows that they must be established with the necessary reliability and precision. Before introducing the SI units, it is instructive to review, briefly, the structure and properties of atoms, which are the building blocks of our world, and indeed the universe, from single-cell living organisms to galaxies. The spectral properties of certain atoms are used in defining the time unit, the second, to extremely high precision.

1.2 ATOMS AND ATOMIC CLOCKS

Atoms are made up of a central nucleus, which is positively charged, surrounded by negatively charged orbiting electrons. Most of the mass of an atom is located in the nucleus. In a neutral atom, the number of electrons is equal to the atomic number Z of protons in the nucleus. The number of uncharged neutrons N in the nucleus of an atom with a particular Z can vary over a small range, giving rise to what are termed isotopes. The atomic mass number A is defined as $A = N + Z$, which is the number of nucleons (protons plus neutrons) in the nucleus. Since the proton and neutron masses are almost the same, and neglecting the very small electron mass contribution, it follows that A is, to a good approximation, equal to the mass of an atom in nucleon mass units. Atoms are identified using an abbreviation of their chemical name together with the mass number A as a superscript. For example, helium atoms with $Z = 2$ and $N = 2$ are identified as ^{4}He.

The lightest atom, hydrogen, is designated as ^{1}H since Z and A are both equal to unity, with the atom consisting of a proton and an orbiting electron. It is interesting to note that hydrogen constitutes roughly 92% of the atoms in the universe. There are two isotopes of hydrogen called deuterium ^{2}H ($Z = 1$ and $A = 2$) and tritium ^{3}H ($Z = 1$ and $A = 3$). Tritium is unstable and undergoes radioactive decay into a helium isotope ^{3}He ($Z = 2$ and $A = 3$). The next lightest atom is ^{4}He ($Z = 2$ and $A = 4$) with two electrons in a neutral atom. Helium makes up close to 8% of the atoms in the universe. In terms of the *mass-fraction* of atoms, hydrogen contributes 74% while helium makes up 24% and heavy elements the remaining 2%. The most massive naturally occurring atom is the uranium isotope ^{238}U ($Z = 92$ and $N = 146$), which is unstable and undergoes radioactive decay with a lifetime comparable to the age of the Earth (~4.5 billion years). The other main uranium isotopes ^{235}U and ^{234}U also undergo successive radioactive decay to lighter elements, as do the more massive elements, which have been produced in experiments using charged particle accelerators. The periodic table of the elements is given in Appendix 2.

It is of interest to note that in recent decades compelling astronomical evidence for the existence of what is termed dark matter has been obtained. Dark matter has not been observed directly and its existence is inferred through gravitational effects on other observable astronomical objects. In addition, the rate of expansion of the universe that astrophysicists have detected suggests that it is necessary to introduce a further mysterious entity called dark energy, which is of dominant importance in determining the effective mass of the universe. Considerable research effort is being devoted to establishing the nature and properties of these intriguing and important constituents of the universe.

Atomic spectroscopy has shown that atoms can absorb or emit electromagnetic radiation at particular wavelengths with corresponding discrete frequencies, ranging from the ultraviolet through the visible spectrum to the infrared and beyond. The sharp spectral features that are observed correspond to discrete changes in the electronic states of the particular atoms whose spectra are being examined. The development of quantum mechanics in the early part of the twentieth century provided the theoretical basis for a deep understanding of atomic scale phenomena including atomic spectra. Since the frequency of electromagnetic radiation can be measured with great precision, it became clear to scientists that atomic transitions could be used in a clock mechanism similar in concept to the use of the frequency of oscillations of a pendulum in a mechanical clock. The current time standard is based on the frequency of a particular transition between close-lying electronic states that occurs in the microwave range for caesium-133 (^{133}Cs) atoms in their ground state. In Cs clocks, microwave radiation from a high precision source is matched to the frequency of the atomic transition which is 9,192,631,770 oscillations per second, or hertz (Hz), the frequency unit named in honour of Heinrich Hertz. The time unit of 1 s is then defined as the time taken

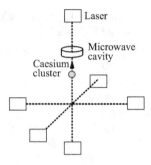

FIGURE 1.1 Schematic representation of the heart of the caesium fountain atomic clock. Six orthogonally aligned laser beams form a trap for a caesium atom cluster. By manipulating the laser beams, the cluster is lobbed upwards through a microwave cavity operating at a selected frequency, The cluster then falls back down through the cavity and is optically examined using a laser beam and a fluorescence detector (not shown) to determine how many of the atoms have undergone microwave-induced transitions. The process is repeated regularly to lock the frequency of the applied microwave clock radiation to that of the caesium atom microwave absorption frequency.

for 9,192,631,770 oscillations. Several high-accuracy caesium clocks have been constructed and are located in countries around the world.

Caesium clocks have evolved over the decades since they were first developed and only the most recently developed and most accurate version will be described here. A sketch of the caesium fountain clock is given in Figure 1.1.

Beams from six orthogonally aligned lasers form a trap for a cluster of caesium atoms, which is suspended at low temperature in a vacuum chamber. By sequentially turning off the laser beams, the caesium cluster is lobbed upwards through a microwave cavity where atomic transitions are induced by microwave radiation. After reaching a maximum height, the cluster falls back down through the cavity. Optical spectroscopy involving another laser and a fluorescence detector is used to determine the number of Cs atoms that have undergone microwave-induced transitions while passing through the cavity. The process is repeated regularly and, using feedback, the microwave clock frequency is kept matched to the Cs transition frequency with high accuracy. The instrument is complex and this brief outline of how it operates does not do it justice. Readers are encouraged to visit the U.S. National Institute for Standards and Technology (NIST) website for further details.

These caesium clocks are accurate to 1 s in 100 million years. Secondary standard caesium clocks are commercially available. International standard or atomic time is based on input from clocks of this kind around the world. Other atomic clocks, which involve ^{87}Rb transitions, are also used for precision timekeeping and have the advantage of being fairly compact. Developments of atomic clocks that operate at optical frequencies, using strontium ions for example, provide orders of magnitude higher precision than the caesium clock.

Atomic clocks in Earth orbiting satellites provide the basis for global positioning system (GPS) technology. Twenty-four satellites, each equipped with four atomic clocks, circle the Earth twice per day in precise orbits that can be adjusted with rocket motors. GPS receivers for travel guidance purposes continuously pick up radio signals from the four satellites that are in view from the point of interest on the Earth's surface. Precise time and position information permit the receiver to calculate its location coordinates to the required accuracy.

1.3 THE SPEED OF LIGHT AND THE UNIT OF LENGTH

The Danish astronomer Ole Rømer's observation and study of the eclipses of the moons of Jupiter in 1676 showed that the speed of light is finite and measurable. Over the years since Rømer's discovery, many increasingly accurate measurements of light's speed have been made using various different experimental approaches. These include time-of-flight experiments, which determine the time taken for a light signal to travel a precisely known distance from a source to a mirror and back, as well as cavity resonators and laser interferometry. From the predictions of classical electromagnetism, derived by James Clerk Maxwell in 1865, together with Albert Einstein's theory of special relativity, published in 1905, it became clear that the speed of light in vacuum, which is denoted by denoted by c, is a fundamental physical constant, which plays a special role in science. By the mid-1970s, the value $c = 2.99792458 \times 10^8$ m/s for light's speed in vacuum was internationally accepted. In 1983, an international conference resolved that c should be *fixed* at the value given above. The meter is then defined using $c = 2.99792458 \times 10^8$ m/s and the caesium clock-based definition of the second. The meter is thus defined as the distance light travels in vacuum in $1/299792458$ s $= 3.335640952 \times 10^{-9}$ s .

Exercise 1.1: Calculate the time it takes for light to travel 1.0 km. As an approximation take $c = 3 \times 10^8$ m/s.
  The time t for light to travel $d = 1.0$ km is given by

$$t = d/c = \left(1 \times 10^3\right)/\left(3 \times 10^8\right) = 3.33 \text{ microseconds.}$$

An advantage of specifying the meter in terms of the *defined* value for the speed of light is that it will not require any change as the precision of measurements increases. In standard laboratories, length calibration is, for convenience, not carried out by measuring the distance light travels in the designated time. It is simpler to use precise laser interferometer techniques involving a stabilized helium–neon (He–Ne) laser beam, whose properties are determined by transitions between the atomic states of Ne atoms in the gas mixture. The laser light beam is in the red region of the visible spectrum and its wavelength λ can be determined with high precision. The value $\lambda = 632.99121258$ nm has been internationally accepted. Thus, the wavelength of the He–Ne laser beam provides a high precision reference standard for length measurement. Figure 1.2 shows a blow-up of a small portion of a laser beam.

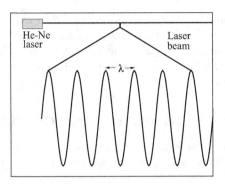

FIGURE 1.2   The upper part of the figure depicts a laser beam from a He–Ne laser, while the lower part shows a highly expanded portion of the beam covering several wavelengths λ.

Exercise 1.2: Determine the number of wavelengths n of a stabilized He–Ne laser beam per meter in vacuum. Use $\lambda = 632.99121258$ nm.

The number of wavelengths per meter is $n = 1/632.99121258 \times 10^{-9} = 157980076.20$, with an uncertainty of parts in 10^{11}.

1.4   THE UNIT OF MASS

The kilogram (kg) is no longer defined as the mass of a platinum–iridium cylinder kept in a vault in the Paris Archives. By international agreement, a new definition of the kg came into effect in May 2019 based on the extremely accurate value obtained for Planck's constant h, which has units of kg m^2 s^{-1}. Planck's constant is of fundamental importance in quantum mechanics. The change in the definition of the kg followed years of effort, which steadily improved the precision of measurements made of h using a special instrument called the Kibble balance. Finally, in 2018, the Commission on Weights and Measures adopted the new definition. A schematic representation of the components at the heart of a modern Kibble balance is given in Figure 1.3.

Just as the speed of light c has an assigned value, Planck's constant has been given a value based on extremely precise measurements. The value adopted is $h = 6.62607015 \times 10^{-34}$ kg m^2/s. Using this fixed value for h, the Kibble balance has become an instrument for determining the mass of an object. An advantage of defining the unit of mass in terms of Planck's constant is that the unit is now linked to the fundamental definitions of time and length. Note that while the Kibble balance is a complex instrument, the basic physics used in determining the mass of an object is not complicated. Firstly, in what is called the weighing mode, an electromagnetic force produced by an applied magnetic field acts on a current-carrying coil, with the current set to balance the gravitational force on the test mass. Secondly, in the velocity mode, the coil is moved in the applied magnetic field at a carefully controlled

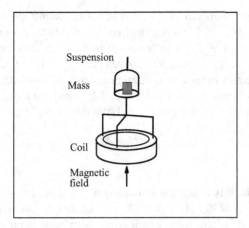

FIGURE 1.3 Schematic depiction of the main components of a modern Kibble balance. The downward gravitational force on the mass shown is balanced by an upward electromagnetic force exerted by the stable magnetic field on the current-carrying coil. The balance operates in high vacuum with high-precision electrical measurements made using special quantum effects that involve Planck's constant.

speed, resulting in an induced voltage in the coil. By combining the results of the two experiments, the mass is obtained. Planck's constant comes into the experimental analysis through the extremely high precision electrical measurements made in the two modes. The measurements are carried out at low temperatures using quantum effects known as the Josephson effect for voltage (velocity mode) and the quantum Hall effect for current (weighing mode). In early experiments, the goal was to measure h with reproducible high precision. After fixing the value of h, the measurements now allow test masses to be determined with high precision.

While there are many other SI units that are important in science, the units of mass, length, and time discussed above are of primary importance in developing classical mechanics. Related units including the unit of force, the newton, are introduced later in the book when needed.

1.5 MASS, WEIGHT, AND NEWTON'S LAW OF UNIVERSAL GRAVITATION

It is important to distinguish between the concepts of mass and weight. Weight is a measure of the force exerted by the Earth's gravitational attraction on a body, while mass is a measure of the nature and composition of a body, which at the microscopic scale is made up of atoms. While the mass of an object can be regarded as fixed, provided Einstein's velocity dependent relativistic effects are small and can be ignored, the weight of an object depends on its location in relation to the Earth's surface. In the Kibble balance measurements, which are described in Section 1.4, it is the weight of a test object that is balanced by an electromagnetic force. Allowance must therefore be made for any variations in the gravitational force, and this is done by carrying out separate calibration measurements.

In dealing with gravitational effects, it is necessary to introduce the unit of force, the newton, denoted N, which in terms of mass, length and time is given by $1\,N = 1\,kg$ $m\,s^{-2}$. The force concept, and Newton's laws of motion, are discussed in detail in later chapters. For the present purposes, it is convenient to simply regard a force as a push or a pull on an object as experienced in everyday life. In particular, the gravitational force F, which acts between two masses m and M separated by a distance r is given by Newton's famous (1686) law of universal gravitation as

$$F = \frac{G\,m\,M}{r^2} \tag{1.1}$$

The constant G, which is called the gravitational constant, has been determined by experiment as $G = 6.67430 \times 10^{-11}$ N m^2/kg^2. Note that the attractive force between two masses falls off as the inverse square of their separation r. Equation (1.1), combined with Newton's laws of motion, can explain the motion of objects near the Earth's surface, including the orbits of artificial satellites and the Moon around the Earth, as well as the orbits of the planets around the Sun. Einstein's 1917 theory of general relativity and subsequent developments, including the recent detection of gravitational waves, have shown that Newton's gravitational law is not a general law applicable to all gravitational effects. Furthermore, the law offers no explanation of how its implied action at a distance operates. However, the law provides an excellent description of the motion of objects under gravitational forces for many situations as discussed later in this book.

Exercise 1.3: The mean radius of the Earth is $R_E = 6.37 \times 10^6$ m. Find the gravitational force on a person of mass $m = 75$ kg standing on the Earth's surface. Assume that the centre of mass of the Earth is effectively concentrated at its centre. (This is a good assumption for a symmetric spherical body.) Take the mass of the Earth as $M_E = 5.97 \times 10^{24}$ kg.
  Newton's gravitational law, given in Equation (1.1), leads to

$$F = \left(6.67 \times 10^{-11} \times 5.97 \times 10^{24} \times 75\right) / \left(6.37 \times 10^6\right)^2 = 736 \text{ N}.$$

Scales for measuring weight are calibrated using a known mass so that they give a reading of the mass of a body in kilograms (or pounds) and not the gravitational force in newtons. For objects at the Earth's surface, Newton's law of universal gravitation can be rewritten as

$$F = m\,g \tag{1.2}$$

with the constant g defined as $g = G\,M_E/R_E^2$. Using the values for the Earth's mean radius R_E and mass M_E given in Exercise 1.3 leads to $g = 9.820$ N/kg or, equivalently, m/s^2. Note that the unit m/s^2 corresponds to an acceleration as introduced in Chapter 2. Thus, a mass in free fall near the Earth's surface experiences an acceleration

g. Variations in R_E occur from place to place on the Earth's surface because the Earth is not exactly spherical, but spheroidal with a slightly larger radius equatorially than that along the polar axis. In addition, variations in altitude and the density of rock substrata near the surface produce variations in g of up to 0.7% around the globe. High-precision weight calibration should therefore be carried out at the site where accurate mass measurements are to be made using sensitive instruments. For simple calculations of the type considered on this book, it is generally sufficient to approximate g as 9.8 N/kg.

1.6 SIZE AND MASS IN THE PHYSICAL WORLD

The ranges of sizes and masses found in nature spans many orders of magnitude as illustrated in Table 1.1 below. Representative values of the height and mass of a person, in SI units, are taken as height 1.7 m and mass 70 kg.

Because of the enormous ranges of sizes and masses of objects in the universe, it is necessary to introduce multiples of the units and subunits specified by the following prefixes: kilo, 10^3; milli, 10^{-3}; micro, 10^{-6}; nano, 10^{-9}; and pico, 10^{-12}. Nuclear and atomic size measurements use the femtometer, fm $= 10^{-15}$ m, and the angstrom, Å $= 10^{-10}$ m, respectively. Astronomical distances are often measured in light years, ly, with 1.0 ly $= 9.46 \times 10^{15}$ m, given by the distance light travels in a year.

1.7 MACROSCOPIC FORCES

There are four fundamental forces in nature called gravitational, electromagnetic, weak, and strong forces. The weak and strong forces are nuclear forces and will not be considered in this book, which focuses on macroscopic phenomena. Having already introduced the gravitational force for the interaction of objects with mass, it is appropriate to consider interactions involving electrically charged objects. The SI unit of charge is the coulomb denoted by C. The charge on an electron is $-1.602176634 \times 10^{-19}$ C, while the proton carries a positive charge $+1.602176634 \times 10^{-19}$ C, which is equal in magnitude but opposite in sign to that of the electron. Macroscopic objects can carry charge, which corresponds to that of many electrons. In dry atmospheric conditions, frictional effects can lead to a build-up of charge on a person's body, and a slight shock will be felt when metal objects are touched.

TABLE 1.1
Ranges of Sizes and Masses Found in Nature

Object	Diameter (m)	Mass (kg)
Proton	1.7×10^{-15}	1.67×10^{-27}
^{1}H atom	1.06×10^{-10}	1.67×10^{-27}
Earth	1.27×10^7	5.97×10^{24}
Sun	1.39×10^9	1.99×10^{30}
Milky Way Galaxy	9×10^{20}	$\sim 10^{42}$

For charged particles such as protons and electrons, and for larger charged objects, the *electrostatic* interaction is governed by Coulomb's law in which the force between two charges q_1 and q_2 separated by a distance r is given by

$$F = k\left(\frac{q_1\,q_2}{r^2}\right) \tag{1.3}$$

Coulomb's law is similar in form to Newton's law of universal gravitation through the inverse square law dependence of the force on the charge separation. In free space, the constant k, which has units N m^2 C^{-2}, is written as $k = 1/4\,\pi\varepsilon_0$. This form for k is chosen for convenience in developing relationships in electromagnetism. The constant $\varepsilon_0 = 8.85418782\times10^{-12}$ C$^2/\left(\text{N m}^2\right)$ is called the permittivity of free space. The value of k in SI units is approximately $k = 8.99\times10^9$ N m^2/C^2. Unlike the gravitational force, which is always attractive, the electrostatic force can be either attractive or repulsive. Like charges repel while unlike charges attract. Electrons and protons have equal and opposite charges, which give rise to attractive forces between these particles in atoms.

For practical reasons, the coulomb is not defined using Coulomb's law. From a measurement point of view, it is convenient to define the unit of electric current, the ampere, which is the rate at which charge passes through an electrical conductor with 1 A = 1 C/s. The ampere is defined in terms of the magnetic force per unit length between two current-carrying conductors separated by a chosen distance. The small discrete charges on the electron and proton, which are given above, are regarded as nature's units.

Exercise 1.4: Determine the attractive force F between an electron and a proton at a separation of 0.05 nm (0.5 Å) corresponding to the radius of the ^{1}H atom. Compare the Coulomb force with the gravitational force, taking the electron mass to be $m_e = 9.1 \times 10^{-31}$ kg and the proton mass to be $m_p = 1.67 \times 10^{-27}$ kg.

For the electrostatic force, Coulomb's law gives

$$F_C = 8.99\times10^9 \times\left(1.60\times10^{-19}\right)^2/\left(0.5\times10^{-10}\right)^2 = 9.2\times10^{-8} \text{ N}.$$

The gravitational force, from Newton's law, is

$$F_G = 6.67\times10^{-11}\times9.1\times10^{-31}\times1.67\times10^{-27}/\left(0.5\times10^{-10}\right)^2 = 4.05\times10^{-47} \text{ N}.$$

The gravitational force is clearly very much weaker, by many orders of magnitude, than the Coulomb force in the hydrogen atom. Gravitational forces therefore play no detectable role in the interactions of particles at the atomic scale.

1.8 FORCES IN THE MACROSCOPIC WORLD

Of the four fundamental forces in nature, it is only the weakest, the gravitational force, which is important on the *macroscopic* scale for objects with mass that are not in physical contact. The nuclear forces are of extremely short range and can be neglected when dealing with assemblies of atoms. Because atoms are electrically neutral, unless they are ionized, long-range Coulomb forces acting on assemblies of atoms are, in general, negligibly weak. Only the gravitational force associated with the Earth's gravitational field produces observable effects that are familiar to all of us.

For objects that are in contact, gravitation will, in many cases, continue to be important but forces at the points of contact come into play. This raises the issue of how the contact forces in the macroscopic world are generated. As an illustrative example, consider two macroscopic objects in close proximity, such as a metal object on a tabletop situated near the Earth's surface. The force of gravity acts on the metal object, causing it to press onto the table. The atoms in the metal try to squeeze into the space occupied by atoms in the table surface, which resist the intrusion. The result is a repulsive interaction between the tabletop and the metal object. Similarly, the table legs stand on the floor and again there is a repulsive interaction between the floor and the table legs. Electromagnetic forces play a role in the repulsive interaction but not in a simple way that is amenable to detailed calculation.

As a second illustration of the action of contact forces, consider a mass suspended on a wire attached to a beam near the Earth's surface. The mass experiences a downward gravitational pull, but the atoms in the wire want to stay together and resist being pulled apart. An upward force in the wire, which is extended by a small but measurable amount, balances the downward force of gravity. Many other examples, including human muscle contraction, can be found in which macroscopic forces are produced by molecular-level interactions.

Later in the book, a clear distinction will be drawn between what are called conservative forces and nonconservative forces. Examples of conservative forces are the gravitational and the Coulomb forces, both of which are associated with fields which permeate space and which can be described by field equations. Nonconservative forces, which include friction and air resistance to projectile motion, are not governed by field equations.

1.9 DIMENSIONAL ANALYSIS

This chapter gives the internationally accepted definitions of the three fundamental physical measurement units which are length [L], time [T], and mass [M]. Other important physical units, called derived units, are expressed in terms of the fundamental units. Examples are the force unit the newton $[M L T^{-2}]$ and the energy unit the joule $[M L^2 T^{-2}]$.

Dimensional analysis provides a simple and useful way of determining the form of relationships that describe physical phenomena. The approach requires that two physical quantities that are related to each other by an equation have the same dimensions. This balance is achieved in general by introducing symbols as exponents that are determined by inspection. A straightforward application to a simple pendulum illustrates the approach.

Exercise 1.5: A simple pendulum consists of a string of length l attached to a support, with a bob of mass m attached to the lower end. The device is situated in the Earth's gravitational field, which gives rise to a downward weight force $m\,g$ acting on the bob. If the bob is displaced and then released, it will execute oscillations with a period t.

In order to obtain an expression for t, it is assumed as a starting point that t is related to the three quantities l, g and m that need to be considered. To simplify the problem, it is assumed that the string is massless. Introducing exponents x, y, and z, the period is written in the form

$$t = l^x g^y m^z \qquad (1.4)$$

In terms of dimensions, $[T] = [L]^x [L\,T^{-2}]^y [M]^z$. Grouping units on the right side of the equation leads to the following values for the three exponents $x = \frac{1}{2}$, $y = -\frac{1}{2}$, and $z = 0$. Inserting these exponent values in Equation (1.4) gives the following expression for the period:

$$t = \sqrt{\frac{l}{g}} \qquad (1.5)$$

Note that the period is determined by the length of the string l and by the gravitational acceleration g but does not depend on the mass of the bob.

While dimensional analysis is useful for providing guidance in describing a variety of physical phenomena, it is necessary to carry out detailed calculations using the laws of physics in order to gain a deep understanding of the subject. The following chapters provide an introduction to the Newtonian mechanics of particles, rigid bodies and fluids, followed by vibrations and waves, and end with a discussion of thermal physics.

2 Motion in Space and Time

2.1 INTRODUCTION

In describing the motion of objects, it is necessary to introduce the basic concepts of displacement, velocity, and acceleration. Any change in these motion variables as a function of time is of particular interest. It is useful to introduce a system of spatial coordinates, called a frame of reference, in which both distances and directions can be shown. An obvious choice in three-dimensional space is a set of Cartesian coordinates with axes x, y, and z chosen along orthogonal direction as shown in Figure 2.1. For certain purposes, and specifically in considering rotational motion, it is preferable to choose spherical polar coordinates, but these will not be used in this chapter.

The displacement, velocity, and acceleration involve both magnitude and direction and are called vector quantities. In dealing with vectors, and specifically with vector addition and subtraction, it is necessary to introduce the elements of vector algebra, and this is done in Section 2.5. Vector multiplication is dealt with in later chapters.

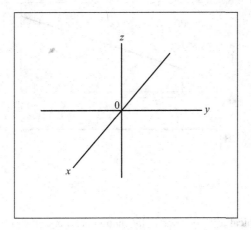

FIGURE 2.1   Representation of Cartesian axes x, y and z in 3D.

DOI: 10.1201/9781003485537-2   **13**

After introducing the variables associated with the motion of objects in space and time, it is instructive to consider motion with *constant* acceleration. In this special case, a set of equations, called the kinematic equations, are found to apply. These equations establish simple and extremely useful relationships between displacement, velocity, and acceleration. Importantly, the kinematic equations hold when the motion of an object is caused by a constant applied force, which produces constant acceleration. The constant force–constant acceleration relationship follows from Newton's laws of motion, which are introduced in Chapter 3. Provided velocities are sufficiently low compared to that of light, so that relativistic effects are unimportant, the kinematic equations satisfactorily describe the observed motion of objects near large astronomical objects, as found in the gravitational field near the Earth's surface. This finding follows from Newton's law of universal gravitation, given in Equation (1.1), together with Newton's second law of motion. In this chapter, the constant acceleration of a falling object in the Earth's gravitational field is taken as an experimental observation.

2.2 MOTION IN ONE DIMENSION

In introducing the displacement, velocity, and acceleration of a moving object, it is convenient, firstly, to consider one-dimensional (1D) motion. Generalization of these results to 2D and 3D follows in a straightforward way using vector notation. Figure 2.2 depicts the position x, velocity v, and acceleration a as a function of time

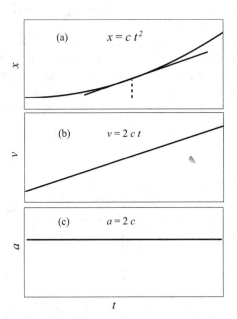

FIGURE 2.2 Graphical representation of the motion of an object with constant acceleration in 1D showing the displacement x in (a), the velocity v in (b), and the acceleration a in (c). The straight line in (a) is drawn as a tangent to the curve and the slope gives the instantaneous speed at the selected time.

t for an object moving with constant acceleration in 1D. The constant acceleration case used in Figure 2.2 facilitates a discussion of the kinematic equations later in this chapter.

Figure 2.2(a), which plots x versus t, shows movement from the initial position x_1 at $t = t_1$ to final position x_2 at $t = t_2$. The displacement is defined as

$$\Delta x = x_2 - x_1 \qquad (2.1)$$

in units of length (e.g. m or km). It is important to distinguish between displacement and distance travelled. If, for example, a runner were to travel to some distant point and then return to her starting point, then her displacement Δx would be zero while the distance traveled would clearly not be zero.

The average velocity of the moving object is given by the rate of change of displacement with time. From Figure 2.2(a) it follows that if in a time interval $\Delta t = t_2 - t_1$ the displacement is $\Delta x = x_2 - x_1$, then the ratio $\Delta x / \Delta t$ gives the average velocity v_{av}. It is often of interest to know the instantaneous velocity, for example, when travelling on a road where speed restrictions apply. Over short intervals Δt, centred at various times t as shown in Figure 2.2(a), it can be seen that the displacement is a function of time written as $\Delta x(t)$. The instantaneous velocity at time t is thus defined as

$$v(t) = \lim_{\Delta t \to 0} \frac{\Delta x(t)}{\Delta t} = \frac{dx(t)}{dt} \qquad (2.2)$$

Equation (2.2) shows that $v(t)$ is simply the first derivative of $x(t)$ with respect to t. Graphically $v(t)$ is the slope of the tangent to the displacement curve at a chosen time t as illustrated in Figure 2.2(a).

The acceleration of a moving object is defined as the rate of change of velocity with time. Just as for velocity, it is necessary to distinguish between the average acceleration a_{av} over a finite time interval Δt and the instantaneous acceleration $a(t)$ at time t. The instantaneous acceleration is defined as

$$a(t) = \lim_{\Delta t \to 0} \frac{\Delta v(t)}{\Delta t} = \frac{dv(t)}{dt} \qquad (2.3)$$

From Equation (2.3), the acceleration $a(t)$ is the first derivative of $v(t)$ with respect to t. Graphically, $a(t)$ is given by the slope of the tangent to the velocity-time curve at a selected time t. Note that from the definitions of $v(t)$ in Equation (2.2) and $a(t)$ in Equation (2.3), it follows that the acceleration is given by the second derivative of $x(t)$ with respect to t:

$$a(t) = \frac{d^2 x(t)}{dt^2} \qquad (2.4)$$

The plot of $v(t)$ versus t in Figure 2.2(b) reveals a straight line with a constant slope, which means that the acceleration is constant as shown in Figure 2.2(c). This particular behaviour is due to the form of the function chosen for the x versus t plot in Figure 2.2(a), whose function is given by $x(t) = x_0 + c\,t^2$, with $x_0 = 0$ for convenience, and c a constant of motion. Differentiation of $x(t)$ with respect to t gives $v(t) = 2c\,t$, corresponding to a linear behaviour of v with t, and further differentiation leads to the time-independent constant value for the acceleration given by $a(t) = 2c$.

In order to generalize the discussion of the motion of objects from 1D to higher dimensions, it is advantageous to introduce vector notation. Displacements in 3D, for example, can have components along the x-, y-, and z-axes in a Cartesian frame of reference. Vector notation and the rules for vector addition and the formation of scalar products are given in Section 2.3. Vector quantities are different from scalar quantities because they are specified by their magnitude and direction, whereas scalars, such as the mass of an object, involve just magnitude.

2.3 VECTORS

2.3.1 Vector Representation

Vectors, such as the displacement of an object, can be represented in a chosen reference frame by an arrow of length proportional to the magnitude of the quantity and pointing in a direction that is related to the physical situation under consideration. Figure 2.3 gives a representation of a displacement of magnitude r at an angle θ with respect to the x-axis in a 2D Cartesian frame. Symbols for vectors are distinguished from those for scalars either by using bold type (e.g. $\mathbf{r}$) or by an arrow above the symbol (e.g. $\vec{r}$).

2.3.2 Unit Vectors

In order to specify the magnitude and direction of a vector in 2D or 3D, it is convenient to make use of unit vectors. In a 3D Cartesian frame, the unit vectors, which have

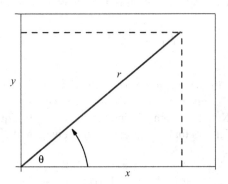

FIGURE 2.3   Graphical representation of a displacement vector $\mathbf{r}$ at an angle θ to the x-axis with magnitude $r = \sqrt{x^2 + y^2}$ and $\tan \theta = y/x$.

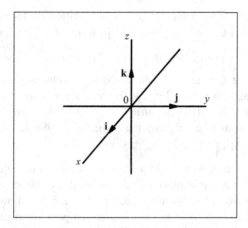

FIGURE 2.4 Unit vectors represented by **i**, **j**, and **k** in a 3D Cartesian frame of reference.

unit length, are directed along the x-, y-, and z-axes, with these vectors represented by the symbols **i** along x, **j** along y, and **k** along z, as given in Figure 2.4. Using unit vector notation, a vector **r** in 3D is written as $\mathbf{r} = r_x\,\mathbf{i} + r_y\,\mathbf{j} + r_z\,\mathbf{k}$ where r_x, r_y, and r_z are the magnitudes of the vector components along the x-, y-, and z-axes. Making use of Pythagoras' theorem, the magnitude of **r** is given by $r = \sqrt{r_x^2 + r_y^2 + r_z^2}$.

Exercise 2.1: In a 2D Cartesian frame of reference, an object is displaced by 8 units of length at an angle of 30° with respect to the x-axis. Express the displacement **r** in terms of unit vectors.

The Cartesian components of vector **r** are given by $r_x = 8\cos 30° = 6.93$ units and $r_y = 8\sin 30° = 4.0$ units. Thus, in terms of unit vectors the displacement is $\mathbf{r} = 6.93\,\mathbf{i} + 4.0\,\mathbf{j}$.

2.3.3 VECTOR ADDITION

In a composite process, involving, for example, two distinct displacements, represented by vector **a** followed by vector **b**, it is necessary to use vector addition to determine the resultant displacement **c**. The sum of the vectors is written as $\mathbf{a} + \mathbf{b} = \mathbf{c}$. Vector **c** can be obtained either by using a geometrical representation of the two vectors **a** and **b** as directed arrows in a coordinate system, or, alternatively, algebraically with the aid of unit vectors. The unit vector method is straightforward and is considered first and designated method 1, while the geometrical approach is method 2.

In method 1, the two vectors **a** and **b** are written in terms of unit vectors as $\mathbf{a} = a_x\mathbf{i} + a_y\mathbf{j} + a_z\mathbf{k}$ and $\mathbf{b} = b_x\mathbf{i} + b_y\mathbf{j} + b_z\mathbf{k}$. The vector sum becomes $\mathbf{c} = \left(a_x + b_x\right)\mathbf{i} + \left(a_y + b_y\right)\mathbf{j} + \left(a_z + b_z\right)\mathbf{k}$ with the Cartesian components of the vectors **a** and **b** along x, y, and z being added separately, and each sum multiplied by the corresponding unit vector. Finally, the resultant **c** follows by summing these components. Vector subtraction is carried out in a similar way with a change of sign from plus to minus in combining the components of **a** and **b**. The resultant is given by $\mathbf{c} = \left(a_x - b_x\right)\mathbf{i} + \left(a_y - b_y\right)\mathbf{j} + \left(a_z - b_z\right)\mathbf{k}$.

Method 2, the geometrical method for vector addition, is illustrated in Figure 2.5(a) and 2.5(b), in which the arrow representation is used for vectors **a** and **b**. There are two equivalent geometrical addition procedures that can be followed. The first procedure involves the triangle rule and the second the parallelogram rule. The use of the triangle rule is depicted in Figure 2.5 (a) in which *one* of the vectors, vector **b**, is displaced parallel to itself until its tail coincides with the tip of vector **a**. Then the resultant vector **c** is given by the arrow drawn from the *tail* of **a** to the *tip* of **b**. The magnitude and direction of **c** can be obtained from measurements made on a scale drawing of the vectors or using trigonometry.

When adding two vectors the triangle rule is a natural choice if, for example, the vectors **a** and **b** represent successive displacements, but it is also applicable when summing vectors such as velocities or accelerations. The parallelogram rule procedure for vector addition is depicted in Figure 2.5(b) and involves displacing one of the vectors parallel to itself until the two *tails* coincide. The parallelogram is then completed by drawing sides parallel to **a** and **b** from the two arrow tips as shown.

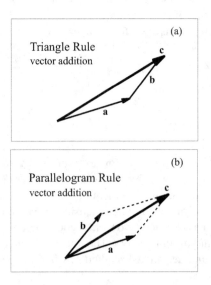

FIGURE 2.5 Addition of the vectors **a** and **b** using (a) the triangle rule and (b) the parallelogram rule.

The resultant vector **c** is given by the diagonal of the parallelogram. Inspection of Figure 2.5(a) and 2.5(b) shows that the triangle and parallelogram vector addition procedures are equivalent. Vector subtraction, given by **a** − **b** = **c**, is carried out by reversing the direction of arrow **b** and then applying the triangle or parallelogram rule to obtain **c**.

Exercise 2.2: An object undergoes two successive displacements in the xy-plane, firstly through a distance of 5.0 m in a direction making an angle of 45° with the x-axis, and secondly through 8.0 m at an angle of 75° with respect to the x-axis. Find the amplitude and direction of the resultant displacement of the object.

Method 1 (unit vectors): The final displacement **r** is given in terms of unit vectors **i** and **j** by $\mathbf{r} = (5\cos 45° + 8\cos 75°)\mathbf{i} + (5\sin 45° + 8\sin 75°)\mathbf{j} = 5.61\,\mathbf{i} + 11.26\,\mathbf{j}$. The distance travelled is obtained using Pythagoras' theorem as $r = 12.6$ m, and the direction, specified by the angle θ which the resultant displacement makes with the x-axis, is given by $\theta = \arctan(11.26/5.61) = 63.5°$.

Method 2 (triangle rule): The two displacement vectors, designated **a** and **b**, are graphically represented in the 2D Cartesian frame in Figure 2.6 with the tail of **b** coinciding with the tip of **a**. The resultant vector **r** is obtained using trigonometry. The cosine rule gives the square of the amplitude as $r^2 = a^2 + b^2 - 2a\,b\cos\phi$ with ϕ the angle subtended by **a** and **b**. Simple geometry gives $\phi = 150°$. Substituting numbers in the cosine rule expression leads to $r^2 = 25 + 64 - 80\cos 150° = 12.6$ m. The displacement direction $\theta = 45° + \psi$ is

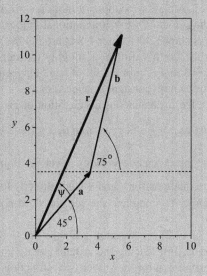

FIGURE 2.6   Addition of vectors **a** and **b** using the triangle rule.

obtained by determining the angle ψ in Figure 2.6 with the aid of the sine rule. This gives $\psi = \arcsin\left(\dfrac{8}{12.6} \times \sin 150^\circ\right) = 18.5^\circ$, and hence $\theta = 63.5^\circ$.

2.3.4 VECTOR MULTIPLICATION: THE SCALAR PRODUCT

Having considered vector addition and subtraction, it is logical, and useful to introduce vector multiplication. There are two types of products that are used in physics, called the scalar product and the vector product, respectively. The scalar product is introduced in this subsection, but the vector product is deferred until later in the book when dealing with torques and rigid body motion.

The scalar product (or dot product) of two vectors $\mathbf{a}$ and $\mathbf{b}$ is defined as $\mathbf{a} \cdot \mathbf{b} = a\,b\cos\theta$ where θ is the angle that vector $\mathbf{a}$ makes with vector $\mathbf{b}$. Note that the product yields a *scalar* outcome. The scalar product involves multiplying the amplitude of $\mathbf{b}$ by the *projected* amplitude of vector $\mathbf{a}$ on $\mathbf{b}$. The scalar product is a maximum for $\theta = 0$ and zero for $\theta = \pi/2$. Scalar products of the unit vectors are readily obtained as $\mathbf{i} \cdot \mathbf{i} = \mathbf{j} \cdot \mathbf{j} = \mathbf{k} \cdot \mathbf{k} = 1$ while $\mathbf{i} \cdot \mathbf{j} = \mathbf{j} \cdot \mathbf{k} = \mathbf{i} \cdot \mathbf{k} = 0$. As an important application, scalar products are used in obtaining the work done by forces in moving their point of application through some distance. Details are given in Chapter 4.

2.4   THE KINEMATIC EQUATIONS

In considering the motion of an object on which a force acts, the important special case of motion with constant acceleration arises when the force acting on the object is constant. This result follows from Newton's second law of motion, which is introduced in Chapter 3. A classic example of motion with constant acceleration is provided by a mass falling in the Earth's gravitational field. Using Newton's law of universal gravitation, it is shown in Chapter 1 that around the globe the acceleration of a falling object near the Earth's surface is roughly the same and, as an approximation, is given by $g = 9.8$ m/s^2. A simple set of equations, known as the kinematic equations, applies to motion with constant acceleration.

From Equation (2.3), the instantaneous acceleration of an object is given by the rate of change of velocity as $a = \dfrac{dv}{dt}$. For *constant a*, it follows that v must be a linear function of time t. The instantaneous velocity given in Equation (2.2) is $v = \dfrac{dx}{dt}$. Integration of the linear equations for a and v leads directly to two of the kinematic equations, as shown below. To complete the set, the third equation is obtained by combining the first two.

Integration of Equation (2.3) expressed as $\int dv = \int a\,dt$, gives $v = a\,t + C$ where C is a constant of integration. Introducing the initial velocity condition $v = v_0$ at $t = 0$ leads to the following expression for the velocity as a function of time,

$$v = v_0 + a\,t \qquad (2.5)$$

This is the first kinematic equation. As required, the velocity increases linearly with time.

Integration of Equation (2.2), using Equation (2.5) to substitute for v, leads to $x = \int v \, dt = \int \left(v_0 + a \, t \right) dt$, which yields $x = v_0 \, t + \frac{1}{2} a \, t^2 + K$ with K a constant of integration. Taking $x = x_0$ at $t = 0$ gives the displacement as $x = x_0 + v_0 \, t + \frac{1}{2} a \, t^2$. Introducing the *net* displacement as $\Delta x = x - x_0$ leads to the second kinematic equation,

$$\Delta x = v_0 \, t + \frac{1}{2} a \, t^2 \tag{2.6}$$

The plots in Figures 2.2(a), (b), and (c) correspond to the constant acceleration case showing respectively the displacement, velocity, and acceleration behaviour with time.

It is useful to combine Equations (2.5) and (2.6) in order to eliminate the time t. This is done by forming the square of both sides of Equation (2.5) and then substituting for terms in t and t^2 using Equation (2.6). This algebraic procedure gives $v^2 = v_0^2 + 2a \, v_0 \, t + a^2 \, t^2 = v_0^2 + 2a \left(v_0 \, t + \frac{1}{2} a \, t^2 \right)$, which simplifies to the third kinematic equation,

$$v^2 = v_0^2 + 2a \, \Delta x \tag{2.7}$$

The three kinematic equations given in Equations (2.5), (2.6), and (2.7) are grouped together below for convenience:

$$v = v_0 + a \, t \tag{2.5}$$

$$\Delta x = v_0 \, t + \frac{1}{2} a \, t^2 \tag{2.6}$$

$$v^2 = v_0^2 + 2a \, \Delta x \tag{2.7}$$

If $x_0 = 0$, then the displacement Δx during the accelerated motion is simply x. In the limiting case, $a = 0$, the equations simplify, and the displacement is $\Delta x = v_0 \, t$ as expected when the velocity is constant.

In summary, Equation (2.5) is a relationship between velocity and the time elapsed since the start of the acceleration process, while Equation (2.7) connects velocity and displacement. Equation (2.6) gives the displacement as a function of time and provides the basis for determining the x-t trajectory of the accelerated motion. Illustrative examples of applications of the kinematic equations are given in the following section for both constant 1D and 2D accelerated motion.

2.5   APPLICATIONS OF THE KINEMATIC EQUATIONS

The constant acceleration kinematic equations (2.5), (2.6), and (2.7) are extremely useful in considering the motion of an object subject to a constant force. Even though motion may occur in more than one dimension, it is often the case that acceleration is associated with one particular direction. For example, for objects moving in the Earth's gravitational field, the downward force of gravity will produce downward acceleration. Since displacement, velocity, and acceleration are vector quantities, care must be taken with signs in writing down equations that describe the motion. If the upward direction is taken as positive, then the gravitational acceleration g is negative since it is directed downwards. When no horizontal forces act on an object, its horizontal component of velocity remains constant. For motion near the Earth's surface, the acceleration due to gravity is taken as $g = 9.8$ m/s^2. The following examples provide illustrative applications of the kinematic equations to the motion of objects subject to constant acceleration in a fixed direction.

2.5.1   DYNAMICS IN 1D

The application of the kinematic equations to the uniformly accelerated motion of an object in 1D is straightforward because the motion is always parallel to the applied force. Exercises 2.3 and 2.4 are illustrative examples of this type of accelerated motion.

Exercise 2.3: A drag race car starts from rest and has an initial acceleration of 8.5 m/s^2. Determine (a) the speed of the car after 3 s and (b) the distance travelled in this time.

(a)   The speed is obtained using Equation (2.5) with $v_0 = 0$. This gives the speed after 3 s as $v = 25.5$ m/s. To convert the speed to km/h, use is made of the conversion factor 1 m/s $= (10^{-3} \times 3600)$ km/hr $= 3.6$ km/h, which leads to $v = 91.8$ km/h.

(b)   Using Equation (2.6), the distance travelled in 3 s is $\Delta x = \dfrac{1}{2} a\, t^2 =$

$\left(\dfrac{8.5}{2}\right) \times 9 = 38.3$ m.

Exercise 2.4(a): A ball is released from rest at a height $h = 2$ m above the floor. How long will it take to reach the floor?

Exercise 2.4(b): If the ball were thrown upwards with an initial speed of 4 m/s from the same initial position, determine how long it would take to reach the highest point of its trajectory. What is the maximum height above the floor reached by the ball, and how long will it take for the ball to reach the floor? Take $g = 9.8$ m/s^2.

(a) Equation (2.6) with $v_0 = 0$ becomes $h = \frac{1}{2}g\,t^2$. The time t for the object to reach the floor is $t = \sqrt{\dfrac{2h}{g}} = \sqrt{\dfrac{2 \times 2}{9.8}} = 0.64$ s. (The negative time solution is unphysical.)

(b) When the ball is thrown upwards, it gradually slows and is instantaneously at rest with $v = 0$ at the highest point of its trajectory. Equation (2.5) for the upward motion becomes $0 = v_0 - g\,t$. Note that the *upward direction* is taken as positive. The time t to reach the highest point is $t = \dfrac{v_0}{g} = \dfrac{4}{9.8} = 0.408$ s.

The upward distance travelled can be obtained using Equation (2.6) in the form $h' = v_0\,t - \dfrac{1}{2}g\,t^2 = 4 \times 0.408 - \left(\dfrac{9.8}{2}\right) \times (0.408)^2 = 0.816$ m. Since the initial launch height is 2 m above the floor, the maximum height reached is $h + h' = 2 + 0.816 = 2.816$ m.

The time t' to reach the floor from the highest point in the trajectory is obtained in a similar way to that used in part (a) by replacing h by $h + h'$. This gives $t = \sqrt{\dfrac{2(h+h')}{g}} = \sqrt{\dfrac{2 \times 2.816}{9.8}} = 0.758$ s. The total time for the upward and downward motion is $t + t' = 1.166$ s.

2.5.2 DYNAMICS IN 2D AND 3D

In analysing projectile motion in 2D or 3D Cartesian reference frames, use is made of the kinematic equations by considering separately the dynamical contributions from the orthogonal spatial directions. Examples are given below. For simplicity, the discussion is limited to motion near the Earth's surface where g is constant. The term projectile motion covers a wide variety of situations in which an object is launched with initial velocity v at an angle θ with respect to the horizontal. In many cases, the Earth's surface can, as an approximation, be assumed to be flat. Note that the situations described as 1D in Exercise 2.4, given above, corresponds to projectile motion for the special case $\theta = \pi/2$. The launch mechanism for projectiles can vary widely. Examples include a kick applied to a soccer ball and the detonation of a charge in a cannon. The launch of rocket ships from Earth is a spectacular example of projectile motion. Allowance must, of course, be made for changes in the gravitational force with altitude if the trajectory takes the craft far from the Earth's surface.

As indicated above, when considering the motion of an object in 2D near the Earth's surface it is convenient to introduce a Cartesian frame of reference with vertical axis y and horizontal axis x. The velocity components are correspondingly v_y and v_x. The trajectory of the object is obtained by applying the second kinematic equation, Equation (2.6), separately to the x and y components of velocity and displacement, making use of the fact that gravity-induced acceleration is limited to the y direction. For convenience, the starting point of the motion is taken as the origin in the reference frame. The constant velocity (zero acceleration) motion parallel to x is described by

$$x = \left(v_0 \cos \theta\right) t \qquad (2.8)$$

where θ is the angle that the initial velocity vector v_0 makes with x at time $t = 0$. Motion parallel to y is governed by

$$y = \left(v_0 \sin \theta\right) t - \frac{1}{2} g\, t^2 \qquad (2.9)$$

with g the gravitational acceleration. Rearranging Equation (2.8) to give $t = \dfrac{x}{\left(v_0 \cos \theta\right)}$, and then substituting for t in Equation (2.9), leads to the following equation for y as a function of x,

$$y = x \tan \theta - \frac{1}{2} g \left(\frac{x}{v_0 \cos \theta}\right)^2 \qquad (2.10)$$

Equation (2.10) describes the object's trajectory. The range of a projectile R is the horizontal distance travelled from launch until it reaches ground level with $y = 0$. Inserting $y = 0$ in Equation (2.10) leads to

$$R = \frac{2v_0^2}{g} \sin \theta \, \cos \theta \qquad (2.11)$$

The range is thus determined by two factors, the initial velocity, and the angle between the horizontal and the launch direction.

Exercise 2.5: At what angle to the horizontal should a projectile be launched from a site on a flat horizontal surface in order to achieve the maximum range for a given initial velocity?

Using the trigonometric identity $2\sin\theta\cos\theta = \sin 2\theta$, Equation (2.11) becomes $R = \dfrac{v_0^2 \sin 2\theta}{g}$. The maximum range R_{max} is reached when

$\dfrac{dR}{d\theta} = \dfrac{2v_0^2 \cos 2\theta}{g} = 0$. It follows that R_{max} is attained when $\theta = \pi/4$. The object should therefore be launched at an angle of $45°$ to the horizontal in order to achieve its maximum range. Figure 2.7 shows the paths followed by an object launched successively at angles of $20°$, $30°$, $45°$, and $60°$ to the horizontal with the same initial velocity of 10 m/s. For launch angles less than $45°$, the travel time to impact with the surface is shorter than for the optimum angle case and this outweighs the higher horizontal velocity component. For angles greater than $45°$, the travel time is longer than at $45°$ but the horizontal component of the object's velocity is reduced compared to that at smaller angles. In the limit of a vertical launch the horizontal travel distance is zero.

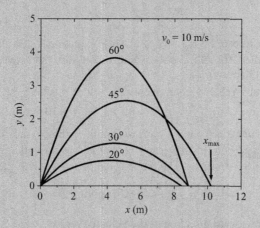

FIGURE 2.7 Trajectories followed by an object launched at different angles to the horizontal with the same initial speed of 10 m/s. The maximum range corresponds to a launch angle of $45°$.

Exercise 2.6: A boy standing on a promontory overlooking a lake throws a pebble in a horizontal direction with initial speed 13 m/s. If the boy's throwing arm is at a height $h = 6$ m above the surface of the lake, determine the horizontal distance the pebble will travel before hitting the water. What is the velocity of the pebble just prior to entering the water? The trajectory of the stone is depicted in Figure 2.8.

The horizontal, x, and vertical, y, motions are independent of each other and are considered separately. Application of Equation (2.6) to the y motion gives $-h = -\dfrac{1}{2}g\,t^2$. The time taken to reach the water surface

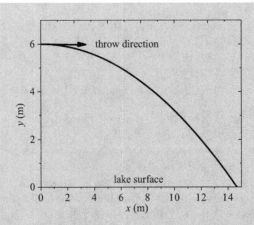

FIGURE 2.8   Trajectory of a stone thrown into a lake. The initial horizontal velocity is 13 m/s at an initial height of 6 m above the surface of the lake.

is $t = \sqrt{2h/g} = \sqrt{2 \times 6/9.8} = 1.107$ s. Using Equation (2.5), the horizontal distance travelled by the stone before hitting the water surface is $x = v_0\, t = 13 \times 1.107 = 14.4$ m.

The downward velocity component just before reaching the water surface is, from Equation (2.5), $v_y = -g\, t = -9.8 \times 1.107 = -10.8$ m/s. The horizontal velocity v_x remains constant at 13 m/s during the pebble's fall. The magnitude of the velocity just before water entry is obtained using Pythagoras's theorem as $v = \sqrt{169 + 118} = 16.9$ m/s. The angle θ that **v** makes with the horizontal is obtained using $\tan\theta = \dfrac{v_y}{v_x} = \dfrac{-10.8}{13} = -0.83$ giving $\theta = -40°$. The minus sign indicates the downward direction of the velocity vector.

While the kinematic equations provide a satisfactory description of many situations involving the accelerated motion of objects, it is necessary to exercise caution when objects such as aircraft or rapid ground transport attain high speeds in the Earth's atmosphere. For situations of this type, it is necessary to allow for air resistance, which becomes increasingly important as the object's velocity increases. For example, objects that fall from a considerable height can reach constant terminal velocities when the downward gravitational force is balanced by the upward air resistance force. The constant acceleration kinematic equations are no longer applicable when the retarding force due to air resistance becomes significant in comparison with the force producing accelerated motion.

Chapter 2 has introduced the concepts of displacement, velocity, and acceleration as vector quantities, which are used to describe the motion of objects through space and time. In Chapter 3 a quantity called the momentum of an object is defined in terms of its mass and its velocity. Momentum plays a central role in describing the dynamics of objects and in formulating the laws of motion called Newton's laws.

3 Momentum and the Laws of Motion

3.1 INTRODUCTION

The behaviour of an object in motion is found to depend on both its velocity and its mass. This conclusion is reached by anyone who has compared the painful experience of being struck by a baseball or cricket ball travelling at speed with the less painful experience of being struck by a tennis ball travelling at roughly the same speed. It is found necessary and useful to introduce the concept of momentum of a moving body in formulating the laws of motion. The momentum of an object is defined as the product of its mass and its velocity. Momentum is a vector quantity through its dependence on velocity. Along with the kinetic energy of a moving mass, which is introduced in Chapter 4, momentum is of fundamental importance in analysing the motion of objects.

The law of momentum conservation, which will be introduced in Section 3.2, is important in considering collision processes over a wide range of mass values from the subatomic, involving fundamental particles such as colliding protons, to the astronomical, such as a meteorite striking a planet. Newton's laws of motion, which are introduced in this chapter, are formulated in terms of the momentum of a body in the presence or absence of external forces. In particular, Newton's famous second law relates the rate of change of momentum of an object to the force acting on the object. Taken together with the law of universal gravitation, Newton's second law provides a description of the motion of the planets and the trajectories of space vehicles exploring the solar system. Note that when dealing with astrophysical events involving very large masses in relatively close proximity, it is necessary to introduce Einstein's general relativity theory, but the classical Newtonian methodology works well for the situations considered in this chapter.

3.2 MOMENTUM AND FRAMES OF REFERENCE

The linear momentum of an object of mass m which is moving with velocity $\mathbf{v}$ with respect to an observer is defined as

$$\mathbf{p} = m\,\mathbf{v} \qquad (3.1)$$

DOI: 10.1201/9781003485537-3

The SI units of momentum are kg m/s. Clearly $\mathbf{p}$ and $\mathbf{v}$ are parallel vectors since m is a scalar quantity. Consider a composite system consisting of N objects, labelled i, with individual masses represented by m_i, each moving with velocity $\mathbf{v}_i$. The total momentum is

$$\mathbf{p} = \sum_i m_i \, \mathbf{v}_i \qquad (3.2)$$

To determine $\mathbf{p}$, the momenta of the individual objects are added using the rules for vector addition.

In specifying the velocity, and hence the momentum, of an object it is necessary to choose a frame of reference. In laboratory situations, it is generally convenient to choose a frame which is fixed with respect to the laboratory floor, which in turn is fixed to the Earth at a particular location. Measurements may, however, be made in a reference frame that is moving with respect to the Earth's surface. Examples include the interiors of ships, aircraft, and orbiting space laboratories. In considering the motion of objects in a moving reference system, it turns out that there is a special class of these frames which are called inertial frames, as discussed below.

Consider two reference frames 1 and 2 that are in relative motion as shown in Figure 3.1. For convenience, the x-axis in frame 1 is chosen to coincide with that in frame 2, which is moving with velocity v in the x-direction with respect to frame 1. From Figure 3.1, it can be seen that the x- and y-coordinates of an object observed in the two reference frames are related as follows: $x_2 = x_1 - v\,t$ and $y_2 = y_1$. The time t is measured from the instant when the origins of the two frames coincide. The relationships between coordinates in the two reference frames constitute what is called the Galilean transformation. Note that times measured by observers in the two reference frames are assumed to be the same. This assumption is valid for $v \ll c$, where c is the speed of light, but breaks down when v approaches c. The Galilean

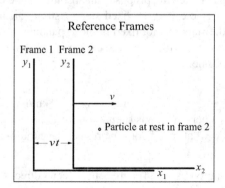

FIGURE 3.1 Two reference frames 1 and 2 in relative motion along their common x-axis, with frame 2 travelling at speed v with respect to frame 1. The coordinates in the two frames are connected as a function of time t by the Galilean transformation provided v is much less than the speed of light.

transformation must then be replaced by the Lorentz transformation used in special relativity theory.

Differentiating the two Galilean coordinate transformation equations with respect to t leads to the Galilean velocity transformation equations, with the x and y velocity components in the two frames related as follows:

$$v_{y1} = v_{y2} \tag{3.3a}$$

$$v_{x1} = v_{x2} + v \tag{3.3b}$$

The y-components of the velocity are the same and equal in the two frames because there is no relative motion along y. The x-components differ by the velocity of frame 2 with respect to frame 1. A particle at rest in frame 2 appears to an observer in frame 1 to be moving with velocity v in the x-direction.

An important insight that is obtained from the Galilean transformation is that there is nothing special that fundamentally distinguishes one Galilean frame of reference from another. It is straightforward to transform velocities from one frame to another provided that the frames are not accelerating. This means that a convenient reference frame can be chosen when considering a particular situation. For example, a reference frame could be chosen in which the object of interest is at rest.

If, in Figure 3.1, the object in frame 2 is not stationary but moving with speed v' at an angle θ with respect to x_2, then the Galilean transformation of velocity components becomes

$$v_{y1} = v_{y2} + v' \sin \theta \tag{3.4a}$$

$$v_{x1} = v_{x2} + v + v' \cos \theta \tag{3.4b}$$

It is interesting to note that Newton proposed an absolute frame of reference that was fixed in relation to what he called the fixed stars. Astronomical observations have, however, shown that the stars are not fixed in the expanding universe, Stars move in galaxies, which in turn move with respect to each other. There is no way to establish an absolute reference frame.

Exercise 3.1: A passenger in a car travelling at 60 km/h along a straight stretch of road sees a person who is running in the opposite direction at 5 m/s on the sidewalk. At what speed does the runner appear to be moving as seen by the passenger in the car?

Adopting the notation used in Figure 3.1, the car's direction of travel on the straight road, alongside the sidewalk, is taken as the common direction of the x-axes for frames of reference attached to the road (frame 1) and to the car (frame 2). No motion occurs along the y direction. In frame 1, the runner moves at speed

5 m/s in the negative x direction. The velocity along x of frame 2 with respect to frame 1 is $v = 60 \times 10^3 / 3600 = 16.7$ m/s. From Equation (3.3b), the passenger in frame 2 obtains the runner's speed as $v_{x2} = v_{x1} - v = -5 - 16.7 = -21.7$ m/s. The minus sign shows that the runner is seen to be moving in the negative x-direction, that is, approaching the car.

In the simple exercise above, it is not made clear how the passenger in the car would measure the speed of the runner. Interestingly, it was the analysis of this type of situation for observers measuring the speed of light signals on a passing train that helped Einstein develop his special theory of relativity.

Exercise 3.2: An object of mass 0.65 kg slides along a smooth horizontal surface at a speed of 12 m/s. What is the linear momentum of the object? What is the momentum in a frame of reference moving at 8 m/s parallel to the direction of travel of the object?

The momentum of the object in the reference frame 1, which is fixed to the smooth surface with the x-axis chosen parallel to the object's path, is $p_{x1} = m\,v_{x1} = 0.65 \times 12 = 7.8$ kg m/s.

In the reference frame 2, which moves at a constant speed of 8 m/s parallel to the path of the object, the speed is given by $v_{x2} = v_{x1} - 8 = 4$ m/s. The transformed momentum in frame 2 is $p_{x2} = 0.65 \times 4 = 2.6$ kg m/s.

3.3   INERTIAL REFERENCE FRAMES

An inertial frame of reference is one in which an object moving at a particular velocity at some instant continues to move with the same fixed velocity as time proceeds. The object could, of course, be stationary as a special case, A necessary condition for the establishment of an inertial frame is that zero net force acts on the object in the chosen frame. Reference frames attached to the Earth's surface are clearly not inertial frames because of the force produced on an object by the Earth's gravitational field. In addition, the Earth is spinning about its axis of rotation, and this results in effects linked to circular motion as discussed in Chapter 5. It is possible to imagine an inertial frame in which an object is situated at a great distance from other masses so that the gravitational forces are vanishingly small. It is a measure of Newton's genius that he could conceive of inertial frames of reference while living in a non-inertial frame. Satellites in orbit around the Earth provide a very good approximation to an inertial frame. These satellites effectively fall towards the Earth, as they move at high speed around the planet, giving rise to the phenomenon of weightlessness. It is important to appreciate that weightlessness does not mean that there is zero gravity. Instead, it means that the object of interest, such as a satellite, is effectively in free fall towards the Earth's centre as it circles the Earth. The object's orbital velocity keeps the satellite moving around the Earth as discussed in Chapter 5. Objects can float motionless if left undisturbed or can move across the spacecraft at constant speed following a slight push on the object by an astronaut. Inertial frames of reference are important in formulating Newton's laws of motion, which are introduced in Section 3.5.

3.4   MOMENTUM CONSERVATION

For a collision involving two or more objects, it has been established that the total momentum of the colliding objects is conserved. This important result is called the law of momentum conservation, which is a fundamental law of physics. The law is expressed mathematically in terms of the changes in the individual momentum vectors for colliding objects i as

$$\Delta \mathbf{p} = \sum_i \Delta \mathbf{p}_i = 0 \qquad (3.5)$$

where $\Delta \mathbf{p}_i$ is the change in momentum of object i during the collision.

In words, the law of momentum conservation states that the vector sum of the changes in momentum for all the objects involved in the collision is zero. Equivalently, the law states that the total momentum of all the objects before the collision is equal to the total momentum after the collision, $\sum_i \mathbf{p}_i^{\text{before}} = \sum_i \mathbf{p}_i^{\text{after}}$. It is important to remember that it is the *vector sum* of the momenta of colliding objects that is conserved.

As an illustration of the use of momentum conservation, consider a collision involving two particles with masses m_1 and m_2, which travel towards each other with velocities $\mathbf{v}_1$ and $\mathbf{v}_2$, respectively, in an inertial frame as depicted in Figure 3.2. After the collision, the particles move apart with velocities $\mathbf{v}_1'$ and $\mathbf{v}_2'$. The law of momentum conservation gives $m_1\left(\mathbf{v}_1' - \mathbf{v}_1\right) = -m_2\left(\mathbf{v}_2' - \mathbf{v}_2\right)$. Note that if $m_2 \gg m_1$, then the change in velocity of particle 1 is much greater than that of particle 2. The inertial frame qualification is introduced so that no external forces act on the particles

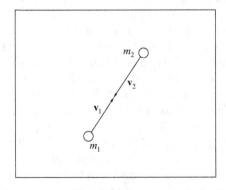

FIGURE 3.2   Collision of two objects of masses m_1 and m_2 which move towards each other with velocities $\mathbf{v}_1$ and $\mathbf{v}_2$ as shown. Momentum is conserved in the ensuing collision.

during the collision. For collisions of high-energy beams of subatomic particles, gravitational effects are negligible and can be ignored in analysing data from collision events. For beams of charged particles, electromagnetic fields are of major importance in determining particle trajectories in a collision event.

Good approximations to inertial frames are achieved in systems that are effectively 2D, such as air tables or ice-skating rinks over which flat objects glide freely. The frictional forces in these systems are very small, permitting momentum conservation to be demonstrated to fair precision in collision processes. These systems are of course 3D, but no motion occurs along directions perpendicular to the supporting surface. The downward force exerted by gravity is exactly balanced by the upward reaction force, which exists between an object and the surface. The thin fluid (gas or liquid) layer at the interface allows the reaction force to be transmitted to the object, but with negligible friction in the surface plane.

Exercise 3.3: A spherical object of mass m_1 travels with velocity $\mathbf{v}_1$ on a low friction air table surface before colliding with a stationary object of mass m_2. If the two objects stick together, determine the velocity $\mathbf{v}_3$ of the combined masses following the collision.

The law of momentum conservation gives $\Delta \mathbf{p} = m_1 \mathbf{v}_1 - \left(m_1 + m_2\right)\mathbf{v}_3 = \mathbf{0}$. Since mass is a scalar, it follows that the two velocity vectors $m_1 \mathbf{v}_1$ and $\left(m_1 + m_2\right)\mathbf{v}_3$ must be parallel to ensure momentum conservation, with

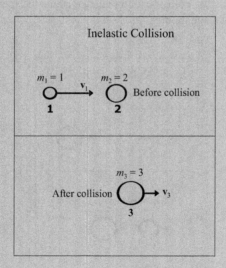

FIGURE 3.3 Inelastic collision of two spherical objects 1 and 2, of masses m_1 and m_2, which can move on a horizontal, low friction table as shown above. Object 2 is initially at rest, while object 1 approaches with velocity $\mathbf{v}_1$. The objects then stick together to form mass $M = m_1 + m_2$, which moves with velocity $\mathbf{v}_3$. Momentum is conserved in the collision. The masses shown are in arbitrary mass units.

$\mathbf{v}_3 = \dfrac{m_1}{m_1 + m_2} \mathbf{v}_1$. The final velocity $\mathbf{v}_3$ is reduced compared to the initial velocity $\mathbf{v}_1$ by the mass ratio $\dfrac{m_1}{m_1 + m_2}$. Using the masses shown in Figure 3.3 gives $\mathbf{v}_3 = \mathbf{v}_1/3$.

Exercise 3.3 provides an example of an *inelastic* collision. Collisions may be either elastic or inelastic, based on the changes in the kinetic energy accompanying a collision. While momentum is *always* conserved in collision events that occur in inertial frames, a scalar quantity called the *kinetic energy* is only conserved in elastic collisions. This distinction between elastic and inelastic collisions is ultimately determined by the nature of the forces that act between the colliding objects as discussed in Chapter 4 in which the important concept of kinetic energy is introduced.

Exercise 3.4: Two ice pucks 1 and 2, with masses m_1 and m_2, respectively, move towards each other with equal speeds v on an ice-rink surface as shown in Figure 3.4. The pucks undergo a head-on collision, and then move apart with their final velocities parallel. If $2\,m_1 = m_2$, obtain expressions for the pucks' speeds v_1 and v_2 after the collision, and then determine the speeds for which the two pucks have the same momentum.

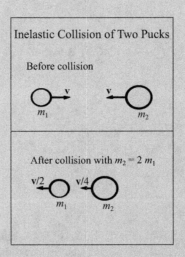

FIGURE 3.4 Two pucks, with masses m_1 and m_2 respectively, travelling towards each other with speeds v on a frictionless surface, undergo a collision as depicted. Momentum is conserved in the collision. The case $m_2 = 2m_1$ is shown.

Momentum conservation gives $m_1 \mathbf{v} + m_2 \mathbf{v} = m_1 \mathbf{v}_1 + m_2 \mathbf{v}_2$. If the velocities are taken as positive in the $+x$ direction, then for the special head-on collision case shown with $m_1 = m$ and $m_2 = 2m$, and accounting for the directions of the velocities before the collision, it follows that $-mv = -m\left(v_1 + 2v_2\right)$. By inspection, the resultant relationship of the speeds $v = v_1 + 2v_2$ is satisfied by, among many others, the values $v_1 = v/2$ and $v_2 = v/4$. In this particular case, the final velocities are parallel and the final momentum in the $-x$ direction is shared *equally* by the pucks.

Note that momentum conservation in two-body collisions leads to a single equation with two unknown final velocities, which in general will not be parallel. For *elastic collisions*, kinetic energy conservation can be used to obtain another equation relating the speeds of the objects. The simultaneous equations provided by momentum and kinetic energy conservation can then be solved to obtain the required final velocities in collision processes.

3.5 NEWTON'S LAWS OF MOTION

The fundamental laws governing the motion of objects are known as Newton's laws. While certain of the ideas were considered by other scientists in the seventeenth century, it was Newton who developed the powerful formalism used in describing the motion of objects. As a striking illustration of his formalism, Newton showed that the laws of motion together with his law of universal gravitation permitted the orbits of the planets about the Sun to be explained in detail. The development of relativity theory and quantum mechanics in the twentieth century has shown that Newton's laws are valid only in the classical limit, but they are extremely useful in describing the motion of objects in a wide variety of situations.

Newton's laws of motion involve the momentum of an object and inertial frames of reference as essential concepts which permit the laws to be stated in compact form as given below.

Newton's first law: An object which is at rest or in uniform motion in an inertial reference frame, and which is not subject to any unbalanced forces, continues at rest or in uniform motion. This is also known as the law of inertia.

In symbols the law states that in an inertial frame $\mathbf{v} = $ constant if $\mathbf{F} = 0$.

Newton's second law: The rate of change of momentum of an object in an inertial reference frame is given by the total applied force acting on the object.

In symbols, the law is written as $\dfrac{d\mathbf{p}}{dt} = \mathbf{F}$.

If the mass of the object remains constant, then the law becomes $m\dfrac{d\mathbf{v}}{dt} = m\,\mathbf{a} = \mathbf{F}$.

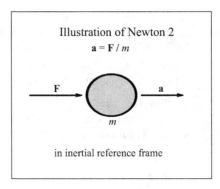

FIGURE 3.5   Newton's second law shows that an applied force **F** acting on an object of mass m in an inertial reference frame produces an acceleration **a** in the direction of the force.

Thus, the second law is conveniently expressed as force equals the mass times the acceleration, or $\mathbf{F} = m\,\mathbf{a}$. This law is of central importance for a range of applications involving the motion of objects subject to applied forces. The SI unit of force, as given in Chapter 1, is the newton with $1\ \mathrm{N} = 1\ \mathrm{kg\ m/s^2}$. Figure 3.5 illustrates the second law for an object of fixed mass.

Newton's third law: For two interacting objects, which are either in contact or interacting via a force field, the force exerted by one object on the other is exactly matched by an equal and opposite reaction force.

In symbols the law states that $\mathbf{F}_{12} = -\mathbf{F}_{21}$ where $\mathbf{F}_{12}$ denotes the force exerted on object 1 by object 2, and $\mathbf{F}_{21}$ is the force exerted on object 2 by object 1.

For two solid objects in contact, the microscopic details of the mechanisms involved in connecting applied forces and reaction forces can be quite complicated. The atoms in the two adjacent surfaces interact as described in Chapter 1. Surface deformation may occur when contact is first made, dependent on the mechanical strengths of the materials. It is only when a state of equilibrium has been reached that action and reaction forces will match as predicted by Newton's third law.

If two objects are not in contact but interact via a force field, such as the gravitational field, then Newton's third law still applies. For example, the gravitational force exerted by the Earth on the Moon is matched by an equal and opposite force exerted by the Moon on the Earth. The orbital motion of the Moon around the Earth prevents the Moon and Earth from crashing into one another. The orbital motion of gravitationally interacting large objects is discussed in Chapter 5.

For convenience, in this chapter the three laws will be referred to as Newton 1, Newton 2, and Newton 3. Note that Newton 1, the law of inertia, is a special case of Newton 2 corresponding to zero net force acting on an object. For a number of applications that are dealt with in this chapter Newton 3 is important when considering the motion of objects on supporting surfaces or hanging from cables.

Newton's laws, together with the kinematic equations, which were introduced in Chapter 2, are very useful in analysing the motion of objects that are acted on by a constant force. Two simple examples are given below.

Exercise 3.5: An electric vehicle of mass 800 kg (including passengers) accelerates from zero to 100 km/h in 6 s along a flat horizontal road. Assuming a constant acceleration, obtain the driving force acting on the vehicle? Ignore wind resistance.

From the first kinematic equation given in Equation (2.5), the speed v of the vehicle as a function of time is given by $v = v_0 + a\,t$. Converting the speed to m/s using 1 km/h $= 1000/3600 = 0.278$ m/s, the constant acceleration is $a = v/t = 100 \times 0.278/6 = 4.6$ m/s^2. Newton 2 gives the average force producing the acceleration as $F = m\,a = 800 \times 4.6 = 3680$ N.

Note that the acceleration is approximately $g/2$. Passengers in the vehicle will feel the push of the seats on their backs while the car is accelerating.

Exercise 3.6: A puck of mass 0.2 kg rests on the horizontal surface of a frictionless air table. If a horizontal force of 0.5 N is applied to the puck, how far will it move in 1.0 s?

From Newton 2, the acceleration of the puck is $a = F/m = 0.5/0.2 = 2.5$ m/s^2. The distance d travelled by the puck is given by the second kinematic equation (Equation (2.6)) as $d = \dfrac{1}{2}a\,t^2 = 1.25$ m.

No motion occurs perpendicular to the air table. The upward reaction force, which balances the weight, is transmitted to the puck through the air cushion.

3.6 APPLICATION OF NEWTON'S LAWS

In applying Newton's laws to the dynamics of an object of mass m located close to the Earth's surface, it is necessary to allow for the downward weight force $F = m\,g$. As discussed in Chapters 1 and 2, the gravitational acceleration is given by $g = G\,M/R^2$. To a good approximation, $g = 9.8$ m/s^2 (or N/kg) using accepted values for G, M, and R.

If an object of interest is supported by a horizontal surface, then Newton 3 shows that the weight force will be matched by an equal and opposite reaction force. For objects that are free to fall under gravity, Newton 2 gives $F = m\,a = m\,g$, and therefore $a = g = 9.8$ m/s^2. The following exercises deal with a variety of situations in which the weight of an object plays a significant role in determining its motion.

An interesting example of the use of Newton 3 involves the apparent weight of a person travelling in an elevator when it is accelerating from rest. Let the passenger's weight be $W_s = M\,g$ as measured on a bathroom scale in the elevator when it is stationary. When the elevator starts to accelerate upwards, the reading on the bathroom scale will increase. This is because the person is also accelerated upwards and the force producing this acceleration is transmitted from the floor of the lift through the scale to the person. In the various force transmission processes Newton 3 plays an

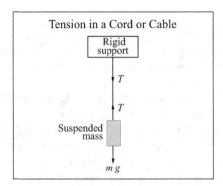

FIGURE 3.6 Mass *m* is suspended on a cord attached to a rigid support. For the mass *m* to be in equilibrium, the upward tension force *T* must be equal to the weight *mg*. The cord transmits the force to the rigid support, which exerts an equal and opposite reaction force on the cord.

essential role by requiring action and reaction to match firstly between the floor of the elevator and the scale and secondly between the scale and the person's feet. The scale reading, which is a measure of the upward reaction force, increases from $W_s = M\,g$ to $W_a = M(g + a)$. When the lift accelerates downwards, the scale reading decreases to give $W_a = M(g - a)$.

Further applications of Newton 3 arise in situations in which a mass is suspended by a cable or cord. The weight of the suspended object $m\,g$ is matched by an upward tension force T in the cable, as illustrated in Figure 3.6. At the point of suspension of the cable, the downward tension force is matched by an upward reaction force exerted by the support.

The tension in a cable or cord can be used to transmit a force between two connected objects that are in motion. Figure 3.7 illustrates a situation of this kind in which a mass m is suspended on a cord, which passes around a frictionless pulley and is then joined to a mass M that rests on a frictionless horizontal surface. The assumption of negligible friction effects is made to simplify the analysis. Friction forces are discussed later in the book.

Application of Newton 2 separately to each of the two masses gives the equations of motion as

$$\text{mass } M: \quad T = M\,a_h \tag{3.6}$$

$$\text{mass } m: \quad m\,g - T = m\,a_v \tag{3.7}$$

The *magnitudes* of the two accelerations a_h (horizontal) and a_v (vertical) in Equations (3.6) and (3.7) must be equal (i.e. $a_h = a_v$) because of the fixed length of the cord connecting them. Combining the two equations by adding them together gives $m\,g = (M + m)a_v$, and hence $a_v = \dfrac{m}{M + m}g$.

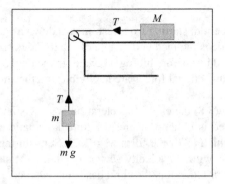

FIGURE 3.7 Masses M and m are connected by an inextensible string, which passes over a frictionless pulley. The gravitational force on mass m produces motion of the coupled masses. The arrows represent the tension T in the string.

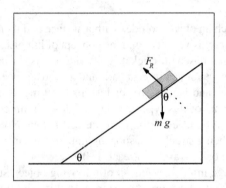

FIGURE 3.8 Accelerated motion of a mass m down a smooth frictionless plane, which is inclined at an angle θ to the horizontal. Geometry shows that the angle between the normal to the plane and the vertical is also θ.

This result could have been written down directly using Newton 2, by taking the force producing the acceleration as $m\,g$ and the total mass moved as $M+m$. The tension force T cancels out in this simple scalar approach. Note that while the mass m is falling with acceleration a_v the tension in the cord is reduced to $T = m\left(g - a_v\right)$.

The motion of an object down a smooth frictionless plane inclined at an angle θ to the horizontal is of interest because the gravitational force on the object has a component $m\,g \sin\theta$ acting down the plane, while a reduced reaction force between the plane and the object $F_R = m\,g\cos\theta$ acts perpendicular to the plane as shown in Figure.3.8.

Newton 2 gives the equation of motion for the sliding object as

$$F = m\,a = m\,g\sin\theta \tag{3.8}$$

The acceleration $a = g\sin\theta$ exhibits a minimum value for $\theta = 0$ and a maximum for $\theta = \pi/2$, as expected.

Exercise 3.7: A child, starting from rest, moves down a water slide of length $L = 4$ m. If the slide is inclined at an angle of $30°$ to the horizontal, what is the velocity of the child just before hitting the water in the pool at the bottom of the slide? Assume that frictional forces are negligible and the slide ends just above the water level.

From Equation (3.8) the child's acceleration is $a = g \sin θ = 9.8 × \sin 30° = 4.9$ m/s². The speed is obtained using the third kinematic equation, given in Equation (2.7), which takes the form $v^2 = 2a\,L$ Substituting values for a and L gives $v = 6.3$ m/s with the velocity vector parallel to the surface of the slide. Note that, as an approximation, the child has been regarded as a compact entity with its mass located at the centre of mass.

3.7 IMPACT

For situations in which an object collides with a surface and then rebounds, it is convenient to introduce impulsive forces and the concept of impulse. Important examples occur in sports such as baseball, cricket, golf, and tennis. In impact collisions, the force that acts between an object (e.g., a ball) and a solid surface (e.g., a baseball bat) is not constant in time and is nonzero for just a short time. It is necessary to adapt Newton's laws to events of this type. The law of momentum conservation does not hold in this inelastic type of collision as discussed below. Note that the change in momentum of a ball being struck by a sporting implement can be large as the change in the direction of the ball's travel can reach $180°$.

Consider the change in momentum $Δp$ of a moving object, such as a squash ball, which strikes a wall that is anchored to the ground. Figure 3.9 illustrates a collision event of this sort. For simplicity, it is assumed that the vertical motion is negligibly small, with the ball travelling at high speed close to horizontally both before and after impact with the wall. As a result of the collision, the velocity of the ball is changed from v_i in the $+x$ direction to v_f in the $-x$ direction. The collision lasts for a very short time, and the details of the process, which involves compression of the ball and slight local elastic deformation of the wall at the point of impact, are not known in any detail. In effect, the wall is considered to be an object of infinite mass.

While the law of momentum conservation cannot be applied to this type of collision, the momentum change $Δp$ of the moving object can be related to a quantity called the impulse I, which is defined as the time integral of the time-dependent force $F(t)$ exerted by the wall on the object during the impact event. This procedure involves, firstly, the use of Newton 2 to determine the instantaneous rate of change of momentum in the collision as $\dfrac{dp}{dt} = F(t)$, and then, secondly, integrating to obtain $Δp = \int_{p_i}^{p_f} dp = \int_0^t F(t) dt = I$. The compact relationship,

$$Δp = I \tag{3.9}$$

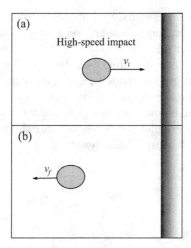

FIGURE 3.9 High-speed collision of a ball with a fixed vertical wall. A time-dependent force acts on the ball during the impact, causing the ball to rebound as shown. Momentum is clearly not conserved in this process.

is useful in dealing with collision events of the type described above. While momentum is obviously not conserved, since $I > 0$ the process may be *quasi-elastic* with only a small change in the kinetic energy, which depends on the square of the speed of the moving object as discussed in Chapter 4. In general, some kinetic energy is transformed during the collision into other forms of energy, specifically sound and heat.

For impact collisions in which the horizontal motion approximation is not applicable, changes in both the vertical and horizontal velocities must be taken into account. Collisions can be investigated experimentally using high-speed photography to follow the motion of the objects and the deformation that occurs during impact.

Exercise 3.8: In a racquetball game, the ball strikes a wall at normal incidence with a speed of 80 km/h. If the ball is in contact with the wall for a time $\Delta t = 20$ ms, and rebounds with a speed of 77 km/h, what is the impulse experienced by the ball in the collision and what is the average force involved? The mass m of a racquetball is 0.04 kg (1.4 ounces).

From Equation (3.9) $\Delta p = p_f - p_i = m\left(v_f - v_i\right) = I$. Choosing the outward normal to the wall as the positive direction, and converting km/h to m/s using 1 km/hr = 0.278 m/s, gives the impulse $I = \Delta p = m(v_f - v_i) = 0.04 \times (21.4 + 22.2) = 1.74$ kg m/s.

The average force is $F_{av} = \dfrac{\Delta p}{\Delta t} = \dfrac{I}{\Delta t} = \dfrac{1.74}{0.02} = 87\,\text{N}.$

The impact concept, which is introduced above, is important in considering any collision process in which a large force acts on an object for a short time. In many sporting activities, a bat or other implement is swung at a ball, and the momentum of the ball is significantly altered. During the collision, a force may be transmitted to the hands of the holder of the bat. The force is found to depend on the position along the bat at which the ball makes impact. In particular, if the ball hits the bat at what is called its centre of percussion, also known informally as the sweet spot, the force transmitted to the hands gripping the bat is minimized. Further discussion of this point is given in Chapter 6 on rigid body dynamics.

4 Work and Mechanical Energy

4.1 INTRODUCTION

When a force acts on a physical system and produces a change in its properties, the force is said to have done work on the system. As an example, consider a system consisting of an object of mass m situated near the Earth's surface and which may be at rest or in motion in the local gravitational field. An applied force can change the velocity of the object and/or its position in the gravitational field, resulting in a change in what is called its mechanical energy. The concept of mechanical energy is of central importance in classical mechanics. As shown below, work and mechanical energy are scalar quantities with SI units joules, abbreviated as J.

It is convenient to distinguish between kinetic energy, which is linked to the motion of an object, and potential energy, which depends on the position of an object in a field such as the gravitational field. Mechanical energy can be conserved in particular situations provided friction effects are negligible. In these circumstances, the sum of the kinetic energy and the potential energy remains constant during a process. An example of this type of process is provided by a falling mass in the Earth's gravitational field provided friction and air resistance effects are negligible. Mechanical energy conservation is a special case of the general law of energy conservation, or, more precisely, mass–energy conservation. This fundamental law covers all forms of energy including mechanical, thermal, chemical, electromagnetic, and nuclear forms.

4.2 MECHANICAL WORK

Consider a force $\mathbf{F}$ acting on an object, which is free to move subject to certain constraints provided by a support. Let the force displace the object through a distance Δs at an angle θ with respect to the force direction as shown in Figure 4.1.

The work W done by the force $\mathbf{F}$ is defined in Equation (4.1) as the scalar product of $\mathbf{F}$ with the object's displacement Δs:

$$W = \mathbf{F} \cdot \Delta \mathbf{s} = F \Delta s \cos \theta \tag{4.1}$$

The work done involves the component of the force parallel to the displacement. The direction of displacement depends upon the constraints on the motion of the object. For

DOI: 10.1201/9781003485537-4

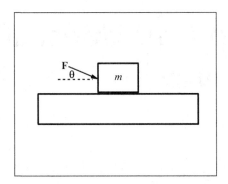

FIGURE 4.1 A force **F** acts at an angle θ to the horizontal on an object of mass m, which rests on a horizontal surface. The block is displaced by a distance Δs along the surface. Work is done by the horizontal component of **F**.

example, when an object is situated on a horizontal surface, no downward motion can occur even though the applied force has a downward component. Newton's third law requires that the net vertical force is zero with upward reaction force equal in magnitude, but opposite in direction, to the sum of the downward forces. In the SI system, the force is measured in newtons, the displacement in meters, and the work done in joules.

For an object of mass m near the surface of the Earth, a downward gravitational force $m\,g$ acts on the object. In discussing the consequences of work done on a system of this type, it is convenient to consider the action of the vertical component of the applied force separately from that of the horizontal component. For a process involving a horizontal force acting on an object situated on a horizontal frictionless surface, the momentum will change with time, corresponding to an increase in velocity in accordance with Newton's second law. The kinetic energy of the object, which is defined below, increases because of the work done by the force. No change in the height of the object occurs, nor is there any change in the vertical component of velocity, which stays at zero in the laboratory frame. If the situation is altered so that the applied force acts vertically upwards, then the height of the object above a chosen reference level, such as the laboratory floor, will change due to the work done by the force. The change in height of the object results in a change in what is called its potential energy, which is introduced below. There may also be an accompanying change in the kinetic energy depending on the direction and strength of the vertical force. If an upward force, which opposes gravity, is only slightly larger than $m\,g$, then the change in kinetic energy will be very small while the height of the object changes, leading to a significant change in potential energy. Expressions for the kinetic energy and potential energy of a body in motion are introduced in the following section.

4.3 MECHANICAL ENERGY

4.3.1 KINETIC ENERGY

Consider a system that consists of an object of mass m that can move freely under the action of an applied force. For example, the object could be in an inertial frame of reference located in an orbiting spacecraft, or, alternatively, earthbound and supported by a frictionless horizontal surface with the applied force acting horizontally. If the force **F** is constant and displaces the object by $\Delta \mathbf{s}$, then Equation (4.1) gives the work done by the force as $W = \mathbf{F} \cdot \Delta \mathbf{s}$. To simplify the discussion, it is convenient to choose the angle θ between **F** and $\Delta \mathbf{s}$ to be zero, so that $W = F\,\Delta s$. From Newton's second law, the constant force produces a constant acceleration of the object, the magnitude of which is given by $a = F/m$.

It is straightforward to relate the work done to the change in the square of the velocity of the object by using the third kinematic equation, Equation (2.7), in the form $v^2 = v_0^2 + 2a\,\Delta s$ Substituting F/m for a gives $m v^2 = m v_0^2 + 2F\,\Delta s$ Replacing $F\,\Delta s$ by W leads to the following result:

$$W = \frac{1}{2}m\left(v^2 - v_0^2\right) \tag{4.2}$$

By adopting the general definition $K = \frac{1}{2}m\,v^2$, Equation (4.2) can be written in the form,

$$W = K_{\mathrm{f}} - K_{\mathrm{i}} = \Delta K = \frac{1}{2}m\left(v_{\mathrm{f}}^2 - v_{\mathrm{i}}^2\right) \tag{4.3}$$

The quantity K is defined as the *kinetic energy* of the moving object. Equation (4.3) shows that the object's kinetic energy is increased by an amount equal to the work done by the applied force. This is an important and useful result. The SI unit of K is joules.

Exercise 4.1: An object of mass 20 kg is free to move on a straight horizontal track and is accelerated from rest to a speed of 4 m/s by an applied force acting parallel to the track. If the distance travelled in the acceleration process is 8 m, determine the change in kinetic energy of the object and the magnitude of the applied force. Ignore the effects of friction.

Since the initial velocity is zero, the change in kinetic energy is $\Delta K = \frac{1}{2}m\,v_{\mathrm{f}}^2 = \frac{1}{2}\times 20 \times 16 = 160$ J. The relationship $W = F\,\Delta s = \Delta K$, with Δs the distance travelled, gives $F = \Delta K/\Delta s = 160/8 = 20$ N.

While physical situations that involve a mass undergoing constant acceleration can be analysed using the kinematic equations, the work–kinetic energy relationship provides the basis for dealing with situations involving the motion of objects subject to variable forces and correspondingly varying accelerations. It is therefore necessary, and also instructive, to obtain Equation (4.3) without making assumptions about the nature of the applied force that leads to a change in the kinetic energy of an object. Details are given below.

It is convenient, initially, to consider motion in 1-D along the x-axis of a frame of reference. The treatment is readily extended to motion in two or three dimensions. The starting point is again Newton's second law, $F = m\, a = m\left(\dfrac{\Delta v}{\Delta t}\right)$, which gives the rate of change of momentum of an object of mass m produced by a force F acting for a time Δt during which the velocity increases by Δv. In a small displacement Δx, the work done by the force is $W = F\,\Delta x$. Substituting for F from Newton's second law leads to $\Delta W = m\left(\dfrac{\Delta v}{\Delta t}\right)\Delta x$. Rearranging gives $\Delta W = m\left(\dfrac{\Delta x}{\Delta t}\right)\Delta v$, and in the limit $\Delta t \to 0$, this results in the equation $dW = m\left(\dfrac{dx}{dt}\right)dv = m\, v\, dv$. Integration of this differential equation gives the required result,

$$W = m\int_{v_i}^{v_f} v\, dv = \frac{1}{2}m\left(v_f^2 - v_i^2\right) \tag{4.4}$$

with v_i and v_f the initial and final velocities along x over the time interval during which the force acts on the mass. Inserting $K = \dfrac{1}{2}m\, v^2$ in Equation (4.4) results in the relationship,

$$W = \frac{1}{2}m\left(v_f^2 - v_i^2\right) = K_f - K_i = \Delta K \tag{4.5}$$

This is the same result as that found using the kinematic equation approach. However, no assumption of a constant applied force is made in obtaining Equation (4.5), which shows that in an inertial frame of reference, the work done by an applied force is equal to the change of kinetic energy of the object to which the force is applied. The derivation of Equation (4.5) is a powerful generalization of the constant acceleration kinematic equation approach. Equation (4.5) thus provides the basis for describing the dynamics of objects subject to forces which vary with time. Defining the change in kinetic energy due to the acceleration of an object as

$$\Delta K = \frac{1}{2}m\left(v_f^2 - v_i^2\right) \tag{4.6}$$

is an important step in developing what is called the work–energy relationship.

In order to lift the restriction that Equation (4.5) is limited to inertial frame situations, it is necessary to broaden the discussion by introducing the potential energy concept to complement that of kinetic energy. The work done by an applied force is, in general, no longer converted entirely into kinetic energy.

4.3.2 POTENTIAL ENERGY

When introducing the concept of potential energy, it is instructive to deal with the special case of gravitational potential energy. Consider a frame of reference in an earthbound laboratory where the Earth's gravitational field gives rise to a constant downward gravitational force $m\,g$ on an object of mass m, with $g = 9.8$ N/kg (or m/s^2). The gravitational force is directed towards the Earth's centre, where the planet's considerable mass is effectively concentrated, when masses experiencing the force are located at, or above, the Earth's surface.

For an object to be at rest in the laboratory frame, it is necessary that the net force on the object be zero. It follows that a vertical force equal in magnitude but opposite in direction to the gravitational force must act on the object. This force could, for example, be the reaction force from a fixed horizontal surface on which the object rests, or the tension in a cord, which is suspended from a fixed support with the free end attached to the object.

Consider an object that is initially at rest in the Earth's gravitational field. If the upward force is increased slightly to $m\,g + \delta F$, the object will gradually increase its height above the laboratory floor from h_i to h_f. The additional force δF does work $\delta W = \delta F\left(h_f - h_i\right)$ and produces a small upward acceleration, and thus a small change δK in the object's kinetic energy. The total work done in increasing the height above the floor is $W = m\,g\left(h_f - h_i\right) + \delta K$. In the limit $\delta F \to 0$, it follows that $\delta K \to 0$ and $W \to m\,g\,\Delta h$ where $\Delta h = \left(h_f - h_i\right)$. In this limit of very small δF, the work done by the upward force has been converted to a form of energy called the potential energy, denoted by U. The potential energy is associated with the height of the object above some reference level, such as the floor, in the non-inertial laboratory frame. The relationship between the work done in the lifting process and the change in potential energy is given by

$$W = m\,g\left(h_f - h_i\right) = U_f - U_i = \Delta U \tag{4.7}$$

Note that while the object was considered to experience a vertical lifting process, it does not matter how the mass is raised from its initial height to its final height. The change in U is always given by $\Delta U = m\,g\left(h_f - h_i\right)$. Any kinetic energy changes that may occur during the raising process simply increase W. The change in potential energy is therefore defined as

$$\Delta U = m\,g\left(h_f - h_i\right) \tag{4.8}$$

with $U_i = m\,g\,h_i$ and $U_f = m\,g\,h_f$.

Potential energy is a stored energy that can be released either slowly by allowing the mass involved to descend gradually while driving some mechanism such as that of a mechanical clock, or rapidly by allowing the mass simply to fall to the floor. In the case of free fall under gravity, the potential energy is converted to kinetic energy in a continuous way. On impact, the object may rebound but will eventually settle on the floor. In the impact process, the kinetic energy is converted into other forms of energy, particularly heat and sound.

Exercise 4.2: Consider an object of mass m initially located at a height h above the floor of an Earth-based laboratory. If the object is allowed to fall, the kinetic energy increases as a function of the vertical distance Δh through which the object has fallen following release. Compare this increase in kinetic energy while the object is in motion with the corresponding decrease in potential energy.

The third kinematic equation, given in Equation (2.7), with initial velocity $v_0 = 0$ gives the square of the downward velocity as a function of the distance Δh through which the object has fallen as $v^2 = 2g\,\Delta h$. Multiplying both sides of this equation by $m/2$ gives the following expression for the increase in kinetic energy $\Delta K = \frac{1}{2}m\,v^2 = m\,g\,\Delta h$.

Since the height of the object above the laboratory floor has *decreased* by Δh, the change in potential energy is given by $\Delta U = -m\,g\,\Delta h$. It follows that $\Delta K = -\Delta U$, which shows that the increase in the kinetic energy of the falling object is exactly equal to the decrease in potential energy. The total energy therefore remains constant during the time that the object is falling in the gravitational field, with potential energy being continuously converted into kinetic energy. This finding is consistent with mechanical energy being conserved during the free fall process.

In the impact with the laboratory floor, the kinetic energy is converted into other forms of energy, including sound waves and heat as mentioned above. The collision raises the temperatures of both the object and the floor locally. Mechanical energy is not conserved in the collision process, although more generally energy in *all* its forms is conserved.

4.3.3 MECHANICAL ENERGY CONSERVATION

The discussion in the preceding subsection shows that the total mechanical energy E of a mass m in the Earth's gravitational field is made up of a potential energy contribution U, associated with its position in the field, and a kinetic energy contribution K, associated with its motion. The total mechanical energy is given by the sum

$$E = K + U \tag{4.9}$$

While Equation (4.9) has been arrived at by considering a specific situation involving a falling object in the Earth's field, the equation is of fundamental importance in mechanics. If no other external forces are present, Equation (4.9) leads directly to the law of mechanical energy conservation, expressed as

$$E = K + U = \text{constant} \tag{4.10}$$

Experiment has shown that the law holds in a wide variety of situations involving what are known as conservative forces. The distinction between conservative and non-conservative forces is explained in Section 4.4. It is shown that gravitational forces are conservative while frictional forces are not. Bearing in mind this limitation on its validity, the law of mechanical energy conservation is extremely useful in solving problems that involve the motion of objects in gravitational fields or in other conservative fields, such as the electric field for charged particles. In the gravitational field case, the moving objects can range from projectiles close to the Earth's surface, assuming air resistance can be neglected, to space vehicles and planets orbiting the Sun. For systems of this type, the total mechanical energy remains constant and therefore changes in K and U sum to zero giving

$$\Delta E = \Delta K + \Delta U = 0 \qquad (4.11)$$

While ΔK and ΔU may vary, they do so in such a way that their sum remains zero.

Exercise 4.3: An object of mass m slides down an inclined plane of length L which makes an angle θ with the horizontal as shown in Figure 4.2. If frictional forces between the sliding object and plane are negligible, find the velocity of the object when it reaches the bottom.

The reaction force between the object and the supporting plane surface acts perpendicular to the plane, and so does not affect the object's sliding motion provided friction is neglected. Mechanical energy conservation, as given in Equation (4.10), is applicable because friction is negligible. The only force of importance is the component of gravitational force acting down the plane. Using energy conservation with $\Delta E = \Delta K + \Delta U = 0$ where $\Delta U = -m\,g\,h = -m\,g\,L\sin\theta$ and $\Delta K = \dfrac{1}{2}m\,v^2$, leads to $v = \sqrt{2\,g\,L\sin\theta}$. The velocity is directed parallel to the plane.

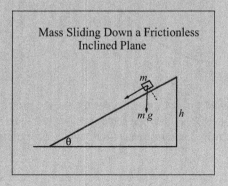

Mass Sliding Down a Frictionless
Inclined Plane

FIGURE 4.2 An object of mass m slides down an inclined plane of length L. Friction forces are taken to be negligible and, to a good approximation, mechanical energy is conserved in the object's descent.

The previous result for the velocity can be obtained using the constant acceleration kinematic equations as shown in Exercise 3.5. However, the next exercise deals with an object travelling down a curved slide along which the force, and therefore the object's acceleration parallel to the surface, change continuously as the object descends. The kinematic equations no longer apply in this case.

Exercise 4.4: An object of mass m slides down a smooth curved surface as shown in Figure 4.3. The top of the slide is at a height h above ground level. Determine the velocity of the object at the bottom of the slide. Assume that friction forces are negligibly small.

Mechanical energy conservation with $\Delta K = -\Delta U$ gives $v = \sqrt{2gh}$, which is the same result as given in Exercise 4.3. This agreement in the values for the velocities at the bottom of the quite different slide geometries comes about because the change in potential energy is the same in the two cases.

Note that the force on the object in the direction of motion at some point in its descent is $F = m\,g\sin\theta$ where θ is the angle that the tangent to the slide makes with the horizontal at the point, as can be inferred from Figure 4.3. In the inclined plane case, θ is constant, and therefore F is constant, while in the curved slide case, F decreases as θ decreases. The kinematic equations cannot be used to describe the motion of an object travelling down a curved slide because the acceleration is continually changing.

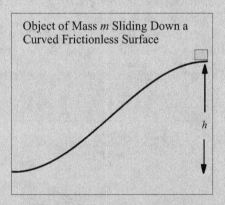

Object of Mass m Sliding Down a Curved Frictionless Surface

h

FIGURE 4.3   An object of mass m slides down a smooth curved slide starting from a position h above floor level. Friction forces are assumed to be very small, and the law of mechanical energy conservation holds to a good approximation for the moving object.

For situations in which the frictional forces between a sliding object and the surface on which it slides are not negligible, it is important to recognize that mechanical energy is no longer conserved. Any work done by frictional forces must be taken into account in describing such a motion as shown later in this chapter. More generally, it is necessary when dealing with the motion of objects produced by various forces to distinguish between conservative forces, such as the gravitational force, and non-conservative forces, such as friction. This distinction is made in the following section.

4.4   CONSERVATIVE AND NON-CONSERVATIVE FORCES

4.4.1   CONSERVATIVE FORCES

Conservative forces are distinguished from non-conservative forces as follows. The work done by a conservative force in moving an object from one point to another is independent of the path followed. In contrast, for non-conservative forces, the work done in moving object does depend on the path followed. Examples of conservative forces are the gravitational force on a mass, and the force on a charged object produced by an electric field. Non-conservative forces include friction forces, which are involved in the relative motion of two surfaces in contact, and viscous drag forces, which act on an object moving through a fluid.

For conservative forces, the work done in moving an object around a closed path is zero, while for non-conservative forces, the work around a similar path is not zero and depends on the path followed. As a simple example of a closed path mechanical work process that involves a conservative force, consider the work done by gravity when an object of mass m is raised in the Earth's gravitational field from floor level to a height h and then returned to its initial position. The work done by the downward gravitational force along the upward path is $W_1 = -m\, g\, h$. The minus sign is introduced because in evaluating the scalar product for the work done by gravity, $W = \mathbf{F}_g \cdot \Delta\mathbf{s}$, the displacement and the gravitational force are anti-parallel. Along the downward return path, the work done by the gravitational force is $W_2 = m\, g\, h$ since the force and the displacement are now parallel. The total work done by the gravitational force in the complete up-down process is $W = W_1 + W_2 = 0$. Note that the focus is on the work done by gravity, and no attention has been paid to the mechanism, or person, supplying the force to raise the object.

Exercise 4.5: Show that the work done by the gravitational force in a process in which a mass m is moved from one position to another in the Earth's gravitational field is independent of the path followed and depends only on the vertical height difference between the initial and final positions.

It is convenient to introduce Cartesian coordinates as shown in Figure 4.4. In terms of unit vectors, the downward gravitational force on the mass m is $\mathbf{F} = -m\, g\, \mathbf{k}$, while an elementary displacement vector has the form $d\mathbf{r} = \mathbf{i}\, dx + \mathbf{j}\, dy + \mathbf{k}\, dz$. The work done by the gravitational force is obtained by evaluating the integral $\int_i^f \mathbf{F} \cdot d\mathbf{r}$ from the initial position i to the final position f as follows:

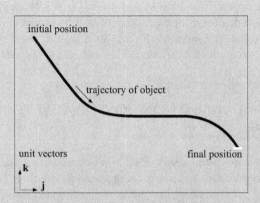

FIGURE 4.4   The sketch shows the 2D trajectory of an object of mass m on which work is done in moving the object from its initial position to its final position.

$$W_{if} = -m\,g\int_i^f \mathbf{k}\cdot\left(\mathbf{i}\,dx + \mathbf{j}\,dy + \mathbf{k}\,dz\right) = -m\,g\int_i^f dz = -m\,g\left(h_f - h_i\right)$$

In the integral, the only scalar product that is nonzero involves the unit vector $\mathbf{k}$. It is apparent that the work done depends only on the vertical height change $h_f - h_i$, with the sign of W dependent on the sign of the height difference.

Note that g is taken as constant in obtaining expressions for the work done in moving the mass m from one position to another. This assumption assumes that the object is never far from the surface of the Earth. In this type of process, it is convenient to choose zero energy arbitrarily to correspond to a reference level such as the laboratory floor or a bench top. A fundamental definition of zero potential energy is necessary in dealing with large separations between the masses. In these cases, it is then convenient to choose zero potential energy to correspond to infinite separation of the masses.

Exercise 4.6: Obtain an expression for the work done by gravity in bringing an object of mass m from a great distance (effectively infinity) to the surface of the Earth. Determine the work done if the object has a mass of 1000 kg.

Using Newton's law of universal gravitation, the attractive force on the mass m at a distance r from the centre of the Earth is $\mathbf{F} = \left(\dfrac{G\,M_E\,m}{r^2}\right)\mathbf{r}$, where M_E is the mass of the Earth, G is the gravitational constant, and $\mathbf{r}$ is a unit vector along the inward radial direction. The work done by the gravitational force is given by

$$W = \int_{\infty}^{R_E} \mathbf{F} \cdot d\mathbf{r} = \int_{\infty}^{R_E} F \, dr \cos\theta = G \, M_E \, m \int_{\infty}^{R_E} \frac{dr}{r^2}$$

$$= -G \, M_E \, m \left[\frac{1}{r}\right]_{\infty}^{R_E} = -G \, M_E \, m/R_E$$

In evaluating the integral, the angle between $\mathbf{F}$ and $d\mathbf{r}$ is $\theta = 0$. For the 1000 kg object, the work done by gravity is

$$W = -\frac{6.67 \times 10^{-11} \times 5.97 \times 10^{24} \times 1000}{6.37 \times 10^6} = -6.25 \times 10^{10} \text{ J.}$$

Note that negative work is done by the gravitational force. The potential energy therefore decreases as r decreases. It follows that a large amount of energy would be required to overcome the Earth's gravitational attraction and transport the mass back to a remote location far from the Earth.

It is useful to introduce the concept of the gravitational potential $U(r)$ associated with the gravitational field produced by a massive object such as a planet or the Sun. This concept is particularly important in dealing with the motion of objects in space, where the law of mechanical energy conservation holds as a very good approximation if small dissipative effects such as ocean tides are neglected. The gravitational potential at a distance r from a massive object is defined as the work done in transporting a test object of mass 1 kg from infinity to the point of interest. The zero of gravitational potential corresponds to the test object being located at infinity. The gravitational potential at a distance r from a mass M is thus given by

$$U(r) = G M \int_{\infty}^{r} \frac{dr}{r^2} = -G M \left[\frac{1}{r}\right]_{\infty}^{r} = -\frac{G M}{r} \tag{4.12}$$

Figure 4.5 shows a plot of the gravitational potential $U(r)$ versus r for the Earth's field. Also shown is the corresponding kinetic energy K of a free-falling 1 kg mass, which, using mechanical energy conservation, is equal in magnitude but opposite in sign to the potential energy. For very large r (approaching infinity), both U and K tend to zero.

While orbiting the Sun, meteoroids and asteroids may encounter a planet such as the Earth. Observations show that many small meteors, with masses of the order of kilograms, burn up in the Earth's atmosphere, producing what are known as shooting stars at night. Asteroids with masses of thousands of kilograms can reach the Earth's surface, giving rise to impact craters. The largest meteor event in recorded history involved the Tunguska meteor, which exploded over Siberia in 1908. The energy of a meteor of this size corresponds to tens of megatons of trinitrotoluene (TNT).

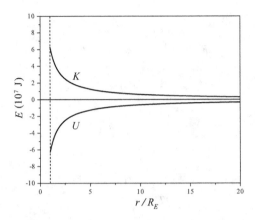

FIGURE 4.5 The gravitational potential U in units of 10^7 J is shown as a function of distance from the Earth's surface in terms of r/R_E where R_E is the Earth's radius. Also shown is the kinetic energy K ($= -U$) of a 1 kg mass approaching Earth from outer space. Small gravitational effects due to other objects far from Earth in the solar system are ignored.

The gravitational potential is a scalar quantity, and contributions to it from several masses can be simply added together. This procedure is easier than adding the gravitational field vectors for a system of several masses. In general, the spatial variation of the gravitational potential $U(x, y, z)$ in 3D can be used to obtain the gravitational field in a given direction by differentiating $U(x, y, z)$ with respect to the corresponding spatial coordinate. A minus sign has to be inserted since the attractive gravitational field increases as the spatial coordinate decreases. For the two-body, case depicted in Figure 4.5, the gravitational field is given by

$$\mathbf{F}_G = -\frac{dU(r)}{dr}\hat{\mathbf{r}} \qquad (4.13)$$

The gravitational field is a vector quantity and is defined as the force per unit mass exerted on a test mass at a point of interest.

Exercise 4.7: Determine the gravitational potential at a point midway between the Earth and the Moon. Use $U(r)$ to obtain an expression for the gravitational field at that point. The Earth's mass is $M_E = 5.972 \times 10^{24}$ kg and that of the Moon is $M_M = 7.35 \times 10^{24}$ kg. The Earth–Moon distance is $R = 3.84 \times 10^8$ and $G = 6.67 \times 10^{-11}$ N m²/kg².

The gravitational potential of the Earth–Moon system at a distance r from the Earth's centre is $U(r) = G\left(\dfrac{M_E}{r} + \dfrac{M_M}{R-r}\right)$ For $r = R/2$, this becomes

$$U(R/2) = \frac{2G}{R}(M_E + M_M).$$

Substituting values gives

$$U(R/2) = \frac{2 \times 6.67 \times 10^{-11}}{3.84 \times 10^8} (6.05 \times 10^{24}) = 2.1 \times 10^6 \text{ J} \cdot$$

Using Equation (4.13), the magnitude of the gravitational field is obtained from

$$F_G(r) = -\frac{dU(r)}{dr} = -G \frac{d}{dr}\left(\frac{M_E}{r} + \frac{M_M}{R-r}\right) = \frac{GM_E}{r^2} - \frac{GM_M}{(R-r)^2}$$

At the midway point, $r = R/2$, this becomes $F_G(R/2) = \frac{4G}{R^2}(M_E - M_M)$. This expression is consistent with the result obtained directly using Newton's law of gravitation for the resultant force on a 1 kg mass produced by two massive bodies.

Exercise 4.8: A meteorite of mass m from deep space approaches the Earth. Estimate the velocity of the meteorite when it enters the Earth's atmosphere.

From Equation (4.12), the gravitational potential is given by $U(r) = -\frac{G M_E}{r}$. Inserting $r = R_E$ as an approximation, and the values for G and M_E from above, gives $U(R_E) = -6.25 \times 10^7$ J as can be seen in Figure 4.5. The kinetic energy per unit mass is thus $K(R_E)/m = 6.25 \times 10^7$ J. Using $K = \frac{1}{2}m v^2$ it follows that $v(R_E) = \sqrt{2K(R_E)/m} = 1.1 \times 10^4$ m/s. This is a very high speed of 4×10^4 km/h, in conventional units.

4.4.2 NON-CONSERVATIVE FORCES

In marked contrast to the path independence of the work done by conservative forces, the work done by non-conservative forces does depend on the path followed when moving an object from an initial position to a final position. Friction provides an important example of a non-conservative force. Friction arises when the surfaces of two objects are in contact, and an applied force acts on one object in an effort to move it across the surface of the other object as shown in Figure 4.6. The magnitude of a friction force, which acts to oppose the motion, depends on the materials involved and the nature of their surfaces. Key surface factors are roughness and the presence or absence of a

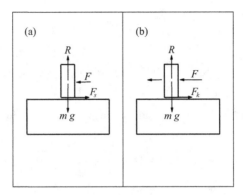

FIGURE 4.6   (a) Friction prevents the upper block of mass m from sliding. The static friction force $F_s = \mu_s m g$ exactly counteracts the applied force F. (b) For $F > \mu_k m g$ the upper block slides over the lower block and the retarding friction force F_k is determined by the kinetic friction coefficient with $F_k = \mu_k m g$. The downward weight $m g$ matches the upward reaction force R.

lubricating fluid between the surfaces. In discussing the work done by friction forces, it is implied that relative motion of an object in contact with another, possibly fixed, surface is occurring and that it is the *dynamic* friction force which is involved. The distinction between static and dynamic friction is dealt with in Section 4.5.

Since friction forces always act to oppose motion, it is clear that the work done against friction depends on the length of the path followed regardless of whether the path is a closed loop or not. At the atomic or molecular level, the interactions that give rise to friction forces are electrical in origin. It is very difficult to make quantitative predictions of friction forces based on microscopic models. An experiment-based, empirical approach involving what are called the laws of friction is generally used in describing friction effects. Further details are given in Section 4.5.

Other non-conservative forces include the viscous drag forces that arise when an object moves through a fluid. In projectile motion, for example, the air resistance drag force always opposes the motion regardless of whether an object is moving up, down or sideways in its path. This is in contrast to the conservative gravitational force, which always acts towards the centre of mass of the object producing the force. It is worth noting for future reference that viscous drag forces depend on the velocity of an object moving through a fluid and become larger and larger as the velocity increases. Since non-conservative forces oppose the motion of a moving object, no part of the work done by the applied force in moving along a closed path is negative. This is different from the conservative force case where the work done in the initial part of a closed trajectory is recovered in a later part of the motion.

4.5   MOTION WITH RETARDING FORCES

The motion of an object produced by a force, such as that due to gravity, may be impeded by a non-conservative retarding force. As mentioned above, examples of retarding forces include friction between solid surfaces in contact and viscous drag on objects moving through fluids. The work done by a retarding force affects the motion and spoils mechanical energy conservation. While the total energy is conserved during motion, it is necessary to allow for the conversion of some mechanical energy into other forms of energy, particularly heat. The macroscopic features of friction and viscous drag forces are discussed in this section.

4.5.1 Friction

Experimental measurements have established what are called the laws of friction. It should be borne in mind that these are empirical laws in contrast to fundamental laws such as Newton's laws of motion. Nevertheless, the laws of friction are very useful in dealing with the mechanics of objects which are acted on by frictional forces. As emphasized in Section 4.4, friction always opposes the relative motion of two objects that are in contact. The two laws of friction are stated as follows:

- *First law:* The friction force opposing the motion of an object across a surface of another object is proportional to the reaction force between the two objects.
- *Second law:* The friction force between two objects is independent of the area of contact.

It is also necessary to distinguish between static and dynamic friction. In static situations, the friction force, denoted F_s increases to match an applied force F that is attempting to slide one object over another. The objects are in static equilibrium under the combined action of the two forces. When the applied force exceeds the limiting maximum value of the static friction force, then sliding motion occurs and it is the kinetic friction force, denoted F_k that now opposes the motion. F_k is, in general, smaller than the maximum value of F_s.

From the first law of friction, it follows that $F_s \leq \mu_s R$ where μ_s is defined as the dimensionless static friction coefficient and R is the reaction force. If the two surfaces in contact are horizontal, then $R = m g$. When $F > \mu_s R$ the situation is altered, and F_s is replaced by the constant kinetic friction force F_k given by $F_k = \mu_k R$ with μ_k the kinetic friction coefficient. Both μ_s and μ_k are typically less than unity, with $\mu_k < \mu_s$. Representative values are $\mu_s = 0.6$ for aluminium on steel, while $\mu_k = 0.9$ for rubber sliding on dry asphalt. Surfaces coated with polymers such as Teflon have very low friction coefficients.

Static friction is depicted in Figure 4.6(a) and kinetic friction in Figure 4.6(b). The lower block shown in Figure 4.6(b) is fixed while the upper block of mass m can move across the lower block.

Exercise 4.9: An aluminium block of mass 12 kg rests on a horizontal steel surface. Determine the maximum horizontal force that can be applied to the block before it starts to move. Obtain a value for the kinetic friction coefficient if the block continues to move at a steady speed following a 20%

reduction in the applied force after the block starts to slide. Take the static friction coefficient for the aluminium steel interface as $\mu_s = 0.6$.

The maximum horizontal force that can be applied before the block slides is $F_{max} = \mu_s \, m \, g = 0.6 \times 12 \times 9.8 = 70.6$ N.

When the block slides at a constant speed, the applied force F is equal to the kinetic friction force F_k. This condition gives $F = F_k = \mu_k \, R = \mu_k \, m \, g$. Putting $F_k = 0.8 \times F_{max} = 56.5$ N, it follows that the kinetic friction coefficient is $\mu_k = 56.5/(12 \times 9.8) = 0.48$. Note that the ratio $\mu_k / \mu_s = 0.8$.

Exercise 4.10: If the steel surface in Exercise 4.9 is tilted at an angle θ to the horizontal, find the maximum angle θ_{max} that can be reached before the aluminium block starts to slide.

The three forces acting on the block are the weight $W = m \, g$ vertically downwards, the reaction force $F_R = m \, g \cos \theta$ perpendicular to the steel plate, and the static friction force $F_f = \mu_s \, m \, g \cos \theta$ acting up the plane as illustrated in Figure 4.7.

The component of the gravitational force acting down the plane is $m \, g \sin \theta$. The maximum angle θ_{max} that can be reached before the block starts to slide corresponds to the magnitude of the friction force up the plane being equal to the component of the weight down the plane, giving $m \, g \sin \theta_{max} - \mu_s \, m \, g \cos \theta_{max} = 0$. This relationship gives $\tan \theta_{max} = \mu_s = 0.6$, and hence $\theta_{max} = 31°$. Note that the friction force which acts up the plane decreases with increasing θ while the component of the block's weight down the plane increases.

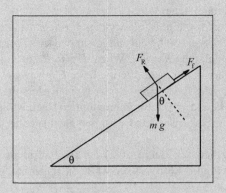

FIGURE 4.7 The aluminium block is positioned on the inclined steel plane, which is canted at an angle θ to the horizontal, and it experiences a frictional force that opposes its downward motion along the plane. For $\theta > \theta_{max}$ the aluminium block slides down the plane, and when $\theta = \theta_{max}$ the forces on the block just balance so that $m \, g \sin \theta_{max} - \mu_s \, m \, g \cos \theta_{max} = 0$.

4.5.2 Viscous Drag

An object moving through a classical fluid experiences a velocity-dependent viscous drag force that impedes the motion. A familiar example is the air resistance experienced when cycling or driving a fast car. In discussing viscous effects in fluids, it is again convenient to adopt an empirical approach based on experiment similar to the approach used in describing friction forces. The representative case considered here deals with air resistance to the motion of an object that falls through the Earth's atmosphere.

The work W_{if} done by the drag force on an object in motion from an initial position i to a final position f depends on the speed of travel and path followed. Experiment shows that the drag force F_d acting on an object that is moving through air near the Earth's surface is proportional to its speed v. This observation leads to the relationship $F_d = C v$ where C is a proportionality constant with SI units of kg/s. If the object of mass m is moving vertically downwards the net force F in the direction of motion is given by $F = m g - F_d = m g - C v$. The falling mass will increase its speed up to a terminal constant value v_t, which is reached when the upward drag force matches the downward gravitational force. This condition means that $m g = C v_t$ and hence

$C = m g/v_t$. Putting $F = m a$ leads to $m a = m g \left(1 - \dfrac{v}{v_t}\right)$ giving $a = g \left(1 - \dfrac{v}{v_t}\right)$. This

expression for a shows that the acceleration decreases steadily towards zero as v tends to v_t. The magnitude of the terminal velocity depends on several factors, including the size and shape of the particular object and the altitude at which the observations are made. These factors are taken into account through the proportionality constant C.

Exercise 4.11: An object drops from a height of 1000 m towards the Earth's surface and experiences a viscous drag force as it falls through the air. Obtain an expression for the ratio of the velocity v to the terminal velocity v_t as a function of time. If the terminal velocity v_t is 28 m/s how long will it take for the object to reach 90% of v_t?

Using the equation for the acceleration $a = g \left(1 - \dfrac{v}{v_t}\right)$ given above and

inserting $a = \dfrac{dv}{dt}$ gives the differential equation $\dfrac{dv}{dt} = g \left(1 - \dfrac{v}{v_t}\right)$. Rearranging

and integrating leads to $\displaystyle\int_0^v \dfrac{dv}{1 - \dfrac{v}{v_t}} = g \int_0^t dt$, and hence $-v_t \ln\left(1 - \dfrac{v}{v_t}\right) = g\,t$.

expressiong antilogarithms this becomes $\dfrac{v}{v_t} = 1 - \exp\left(-\dfrac{g}{v_t}t\right)$ which is the required expression.

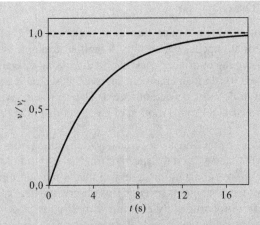

FIGURE 4.8   The plot shows the time dependence of the ratio of the velocity v of a falling object, which is subject to air resistance, to its terminal velocity v_t (= 28 m/s). The exponential growth curve illustrates how a falling body approaches its terminal velocity following its release from rest in the atmosphere above the Earth's surface.

This exponential growth form determines how the speed of the falling object tends to v_t. Inserting the given value for v_t leads to $0.9 = 1 - \exp\left(-\dfrac{9.8}{28}t\right) = 1 - \exp(-0.35t)$. Solving for t shows that the time to reach 90 % of the terminal velocity is 6.6 s. Figure 4.8 gives a plot of v/v_t versus t for the falling object.

4.6   ENERGY CONSERVATION AND NON-CONSERVATIVE FORCES

As discussed in Section 4.4, the law of mechanical energy conservation given in Equation (4.10) holds for systems in which a conservative force, such as the gravitational force, produces motion of an object. Mechanical energy is not conserved when friction or some other non-conservative force is involved. To illustrate this point, consider a block moving under gravity down an inclined plane of length L, which makes an angle θ with the horizontal as discussed in Exercise 4.10. The work done by the component of the gravitational force down the plane is $W_g = m\,g\,L\sin\theta = m\,g\,h$ while the *negative* work done by the friction force acting *up* the plane is given by $W_f = -\mu_k\,m\,g\,L\cos\theta = -\mu_k\,m\,g\,h\cot\theta$. The change in kinetic energy, or the net work done by the combination of gravitational and frictional forces in moving the block down the plane is $\Delta K = W_g - W_f = m\,g\,h\left(1 - \mu_k\cot\theta\right)$. The change in the gravitational potential energy is simply $\Delta U = W_g = m\,g\,h$. It follows that $\Delta K < \Delta U$ with a fraction of the gravitational potential energy being converted into heat and not into kinetic energy of the sliding block.

A striking example of non-conservation of mechanical energy involves the sliding motion of an object at constant velocity down an inclined plane, which makes a carefully selected angle with the horizontal. Under the combined action of the conservative gravitational force component down the plane and the equal and opposite non-conservative kinetic friction force, which acts up the plane, the object moves at a steady speed once motion is started with a small push. No change in the kinetic energy of the object occurs and all the work done by gravity is converted into other forms of energy such as heat. While energy in all its forms is conserved, the mechanical energy, which is made up of potential energy and kinetic energy, is not. The *general* law of energy conservation for this type of process then has the form,

$$\Delta K = -\Delta U - |W_f| \qquad (4.14)$$

The magnitude of W_f and the minus sign are used in Equation (4.14) to emphasize that retarding forces such as friction do negative work on a moving object.

As emphasized above, the great importance of the law of mechanical energy conservation is its general applicability to objects in motion under the action of conservative forces. For example, it is easy to analyse the motion of an object which moves on a curved path along which potential energy is converted to kinetic energy. This feature is lost when non-conservative forces are involved. To illustrate the point, it is instructive to consider the motion of an object down a curved slide firstly with zero friction and secondly allowing for friction. This is done in Exercise 4.12.

Exercise 4.12: A curved slide is shaped in the form of a quadrant of a circle of radius R as shown in Figure 4.9. If a small object, represented by a block, is released from rest at the top of the slide find the object's velocity at the bottom of the slide, firstly by ignoring friction, and secondly by allowing for kinetic friction.

In the case of zero friction, mechanical energy of the sliding object is conserved. The condition $\Delta E = \Delta K + \Delta U = 0$ leads to the relationship $\frac{1}{2}m\,v^2 - m\,g\,R = 0$. The speed of the object at the bottom of the slide is given by $v = \sqrt{2g\,R}$ along the horizontal direction. The change in kinetic energy is given by $\Delta K = \frac{1}{2}m\,v^2 = m\,g\,R$.

When the friction force is non-zero, mechanical energy conservation no longer holds. With allowance for W_f, the work done by the varying friction force, Equation (4.14) applies with $\Delta K = -\Delta U - |W_f|$. The determination of W_f involves summing infinitesimal contributions $dW_f = -\mu_k\,m\,g\cos\theta\,dL$ from a succession of intervals dL down the slide. The angle θ is the angle that a tangent

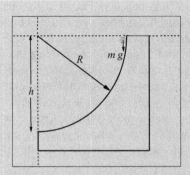

FIGURE 4.9   A small object of mass m moves down a curved slide, made in the form of a quadrant of a circle, and effectively falls through a vertical height $h = R$. The velocity at the bottom depends on a position dependent kinetic friction force which opposes the motion. It is assumed that the flexible object always makes good contact with the curved surface.

to the curved slide makes with the horizontal direction at a given point. Putting $dL = -R\,d\theta$, and converting the sum to an integral, leads to

$$W_f = \mu_k\, m\, g \int_{\pi/2}^{0} \cos\theta\, R\, d\theta = \mu_k\, m\, g\, R\left[\sin\theta\right]_{\pi/2}^{0} = -\mu_k\, m\, g\, R$$

As expected, W_f is negative. The change in kinetic energy is therefore

$\Delta K = \dfrac{1}{2}m\, v^2 = m\, g\, R\left(1 - \mu_k\right)$ and the speed at the bottom of the slide is

$v = \sqrt{2g\, R\left(1 - \mu_k\right)}$. This result shows that the final speed is reduced by a factor $\sqrt{\left(1 - \mu_k\right)}$ compared to the case of zero friction.

Note that the choice of shape, the quadrant of a circle, for the slide is made to simplify the integral. The point of this exercise is to demonstrate that calculations based on Equation (4.14) are, in general, not simple to carry out. Numerical methods may be necessary in such cases.

5 Circular Motion

5.1 INTRODUCTION

Circular motion is concerned with the motion of objects which are subject to a central force constraint. Examples include planetary motion about the Sun and satellite motion about the Earth. In these cases, the orbits may not be exactly circular but rather elliptical as discussed below. The perfectly circular motion of a mass which is attached by a cord to a support on a flat horizontal frictionless table provides an easily visualized model system for analysing this type of motion.

In dealing with the dynamics of circular motion of an object, it is necessary to introduce angular variables including angular velocity and angular acceleration. It is then straightforward to obtain the modified kinematic equations for constant angular acceleration situations. The introduction of the concept of angular momentum and the generalization of Newton's laws to angular motion provides the basis for understanding the properties of rotating systems. The results obtained are of importance in many branches of science and engineering.

5.2 ANGULAR VARIABLES FOR ROTATIONAL MOTION

In order to describe rotational motion, it is natural to introduce angular coordinates. Consider an object undergoing circular motion about a fixed point which is taken as the origin of a set of Cartesian axes. The object's position is specified by the angle θ between a reference direction, which is labelled as a coordinate axis, and the line drawn from the centre of the circular path to the position of the object as illustrated in Figure 5.1. Note that θ is chosen to be the polar angle that the radius r makes with the y-axis. The angle is measured in radians, with 1 radian corresponding to the angle subtended by an arc of length equal to the radius of the object's circular path. In degrees, 1 radian is $\left(r/2\pi r\right)\times 360° = 57.3°$. In Figure 5.1, the angle $\theta = 1$ radian would correspond to the arc length PQ being equal to the radius r. It follows that in general $\theta = s/r$ with s the arc length PQ. The angular displacement over a time interval Δt is given by $\Delta\theta = \theta_f - \theta_i$ where θ_i and θ_f are the initial and final orientations, respectively. The arc length s traversed by the moving object is $s = r\,\Delta\theta$.

DOI: 10.1201/9781003485537-5

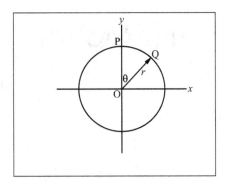

FIGURE 5.1   Coordinates r and θ that are used to describe circular motion of an object about a fixed point at the origin O. The arc length of the segment PQ is given by $s = r\theta$ with θ in radians.

The instantaneous angular velocity of an object executing circular motion is defined as $\omega = \lim\limits_{\Delta t \to 0} \dfrac{\Delta\theta}{\Delta t} = \dfrac{d\theta}{dt}$ and is measured in radians/s. Note that radians are dimensionless. The instantaneous *velocity* of the object is given by $v = r\left(\dfrac{d\theta}{dt}\right) = \omega\, r$ and is represented by an arrow drawn as a tangent to the circular path at the point of interest. Similarly, the instantaneous angular *acceleration* is $\alpha = \dfrac{d\omega}{dt}$. For constant α, the circular motion is described by the rotational kinematic equations given in Section 5.3.

Exercise 5.1: A cyclist speeds around a circular cycle track of radius 60 m covering 20 laps in 10 minutes. Determine the average angular velocity and the average speed of the cyclist.

The average angular velocity is $\omega = \dfrac{\Delta\theta}{\Delta t} = \dfrac{20 \times 2\pi}{600} = 0.21\,\text{rad/s}$. Note that radians are abbreviated as rad. The average speed is $v = \omega\, r = 0.21 \times 60 = 12.6\,\text{m/s}$.

5.3   ROTATIONAL KINEMATICS

The derivation of the kinematic equations for circular motion with constant angular acceleration α follows similar mathematical steps to those used to obtain the linear kinematic equations discussed in Chapter 2. Integration of the angular acceleration equation $\alpha = \dfrac{d\omega}{dt}$ after it is rewritten as $d\omega = \alpha\, dt$ gives the first rotational kinematic equation:

$$\omega = \omega_0 + \alpha\, t \tag{5.1}$$

where ω_0 is the angular velocity at $t = 0$. Integration of Equation (5.1) leads to the second rotational kinematic equation,

$$\theta = \omega_0 t + \frac{1}{2} \alpha t^2 \tag{5.2}$$

The third equation is obtained by squaring both sides of Equation (5.1), and then using Equation (5.2) multiplied by 2α to eliminate t. This procedure results in the third kinematic equation,

$$\omega^2 = \omega_0^2 + 2\alpha \theta \tag{5.3}$$

The forms of the three rotational kinematic equations are similar to the linear kinematic equations, with α, ω and θ replacing a, v and x respectively.

Exercise 5.2: If the cyclist of Exercise 5.1 accelerates uniformly for 5 s and increases the angular velocity of the cycle by 0.05 rad/s, what is the angular acceleration? What is the angular displacement in this period? What distance on the track does the cyclist cover while accelerating?

Rearranging Equation (5.1) leads to the following expression for the angular acceleration $\alpha = (\omega - \omega_0)/t = 0.05/5 = 0.01\,\text{rad/s}^2$. Equation (5.2) then gives the angular displacement during the 5 s period as $\theta = \omega_0 t + \frac{1}{2} \alpha t^2 = 0.21 \times 5 + \frac{1}{2} \times 0.01 \times 25 = 1.17\,\text{rad}$. The distance covered while accelerating is $s = r\,\theta = 60 \times 1.17 = 70.2\,\text{m}$.

5.4 CENTRIPETAL ACCELERATION

An object in circular motion with constant angular velocity ω about a fixed centre point experiences constant acceleration towards the centre. While the magnitude of the velocity remains constant the direction of the velocity vector is continually changing. The constant acceleration is produced by a constant central attractive force such as the gravitational attraction experienced by an Earth satellite or the tension in a cord attached to a whirling object. The acceleration is called the centripetal acceleration. As shown below, the acceleration depends on the square of the angular velocity of the object and on the radius of the circular path.

Figure 5.2(a) depicts an object moving with angular velocity ω in a counterclockwise circular orbit of radius r about fixed centre point O. The arrows labelled $\mathbf{v}_a$ and $\mathbf{v}_b$, which are drawn as tangents to the circle at points a and b, represent velocity vectors, each with length proportional to $v = \omega r$, at two instants separated in time by Δt, which is chosen to be small compared to the time T taken for a complete orbit. The angular displacement of $\mathbf{v}_a$ with respect to $\mathbf{v}_b$ is $\theta = \omega \Delta t$. The vector diagram

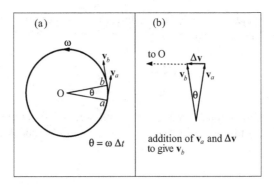

FIGURE 5.2 (a) The circular path depicts an object undergoing circular motion with angular velocity ω. Instantaneous velocities $\mathbf{v}_a$ and $\mathbf{v}_b$ at points a and b are shown for times t and $t + \Delta t$. (b) The plot shows that vector addition of $\mathbf{v}_a$ and the incremental velocity change Δv gives $\mathbf{v}_b$. The centripetal acceleration $\mathbf{a}_C = \Delta \mathbf{v}/\Delta t$ for $\Delta t \rightarrow 0$ is directed towards O at the centre of the circle.

in Figure 5.2(b) shows that over the short time Δt the velocity $\mathbf{v}_a$ is altered by the addition of a vector of magnitude $\Delta v \approx v \, \theta$, which is directed along the radial direction towards O, giving $\mathbf{v}_b$. The corresponding radial acceleration, directed towards O, is the centripetal acceleration with magnitude $a_C = \lim\limits_{\Delta t \to 0} \dfrac{\Delta v}{\Delta t} = v \dfrac{d\theta}{dt} = v \, \omega$. Using $v = \omega \, r$ leads to the following relationship between a_C and ω

$$a_C = \omega^2 \, r \tag{5.4}$$

Equation (5.4) is an important result in describing circular motion. Note that the relationship can be written in terms of the speed as $a = v^2/r$.

 The application of Equation (5.4) to the motion of satellites around the Earth, or to the orbits of planets around the Sun, involves using the gravitational force F_G and centripetal acceleration. While planetary orbits are in general elliptical, some of those orbits, including that of the Earth, are close to circular. These cases can be modelled by considering a small mass m executing circular motion around a large mass M in an orbit of radius r. Newton's second law together with the law of universal gravitation leads to $F_G = \dfrac{G M m}{r^2} = m \, \omega^2 \, r$. This result gives the following expression for the angular velocity:

$$\omega = \sqrt{\dfrac{G M}{r^3}} \tag{5.5}$$

The time for a complete circular orbit of the satellite or planet is $T = 2\pi/\omega$.

Exercise 5.3: A satellite orbits the Earth at a low altitude of 250 km. Determine the angular velocity and the speed of the satellite. Take $M_E = 5.97 \times 10^{24}$ kg and $R_E = 6.37 \times 10^3$ km.

Using Equation (5.5), the angular velocity is given by $\omega = \sqrt{\dfrac{GM_E}{R_E^3}} = \sqrt{\dfrac{(6.67\times10^{-11})\times(5.97\times10^{24})}{(6.37\times10^6)^3}} = 1.24\times10^{-3}$ rad/s. The speed is $v = \omega R = (1.24 \times 10^{-3}) \times (6.37 \times 10^6) = 7.9 \times 10^3$ m/s.

A further discussion of Earth satellites is provided in Section 5.10.

5.5   ANGULAR MOMENTUM

While linear momentum is clearly not conserved for an object undergoing circular motion, because the velocity is continuously changing direction under the influence of a central force, there is a quantity called the angular momentum, which is conserved in this situation. For a mass m undergoing circular motion with radius r and constant tangential speed v about a central point O, the *magnitude* of the angular momentum about O, denoted by l, is taken as $l = mvr = m\omega r^2$. Note that the vectors **r** and **v** are perpendicular to one another. For reasons that will become clear, the angular momentum is defined as a vector quantity. In contrast to the linear momentum vector, **p** = m**v**, which changes with time in circular motion as the velocity changes direction, the angular momentum vector **l** is a constant of motion as discussed in Section 5.6.

The SI units for angular momentum are kg m^2 s^{-1} or, more conveniently for many purposes, J s. In macroscopic rotating systems, the angular momentum can take on a continuous distribution of values. This is no longer the case at the atomic scale where angular momentum is quantized in terms of Planck's constant $h = 6.63\times10^{-34}$ J s, leading to a discrete set of quantum states. The famous Bohr model of the atom, which involves quantization of angular momentum, can explain the spectroscopic feature of light emitted by hydrogen gas in a discharge tube. While this simple model breaks down for heavier atoms, the quantization ideas played a key role in the subsequent development of quantum mechanics.

It is useful to consider a generalized definition of **l**, which allows for situations in which the vectors **v** and **r** are not mutually perpendicular but subtend an angle θ as shown in Figure 5.3. This situation arises, for example, when the angular momentum is required about a point that is not at the centre of rotational motion. Note that in Figure 5.3 the vectors are drawn so that their tails coincide. In the generalized definition of angular momentum about some chosen point, it is the component of **v** perpendicular to **r** that is important, and the magnitude of the angular momentum takes the form $l = mvr\sin\theta$. It follows that l increases from zero for $\theta = 0$ to a maximum value mvr for $\theta = \pi/2$.

Angular momentum exhibits vector properties that are quite different from those of the linear momentum **p** = m**v** of a mass m moving with velocity **v** in a fixed direction. In the case of angular momentum, it is necessary to consider the *product* of two

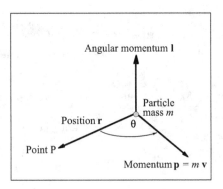

FIGURE 5.3   The angular momentum of mass m about point **P** is $l = mvr\sin\theta$ and is represented by a vector directed perpendicular to the plane containing the position and momentum vectors **r** and **p**.

vectors. This is done using vector product notation and rules, which are introduced in the following section.

5.6   THE VECTOR PRODUCT AND ANGULAR MOMENTUM

As introduced in Chapter 2, the scalar product of two vectors **a** and **b** that subtend an angle θ is defined as $\mathbf{a}\cdot\mathbf{b} = c = ab\cos\theta$. This form gives a scalar result, which involves the product of the magnitude of the component of vector **a** projected onto **b** with the magnitude of **b**. The same result is obtained using the magnitude of the component of **b** projected onto **a** with the magnitude of **a**. The scalar product is useful, for example, in calculating the work done by a force that displaces an object along a direction that is not parallel to the force.

The vector product, or cross product, of vectors **a** and **b** is defined as $\mathbf{c} = \mathbf{a}\times\mathbf{b}$ where **c** is a vector of magnitude $ab\sin\theta$ oriented perpendicular to the plane containing **a** and **b** as shown in Figure 5.4. The magnitude is given by the *perpendicular* component of **a** with respect to **b** multiplied by the magnitude of **b**. The vector product is useful in many situations and, in particular, when discussing angular momentum. The following rule, known as the right-hand rule, is used in establishing the direction of vector **c**. If the fingers of the right hand are curled in the rotation sense (clockwise or counterclockwise) in turning from **a** to **b**, then the direction of **c** is given by the direction of the thumb. For the vectors **a** and **b** depicted in Figure 5.4, the fingers will curl counterclockwise (as indicated by the curved arrow) in rotating from **a** to **b** with the thumb pointing upwards. The resultant vector **c** is aligned along the $+z$ direction. Interchanging the order of vectors in the vector product would change the sense in which the fingers of the right-hand curl, from counterclockwise to clockwise, with the thumb pointing downwards. It is convenient to introduce the cross-products of the unit vectors in 3D. Using the definition of the cross-product gives $\mathbf{i}\times\mathbf{i} = \mathbf{j}\times\mathbf{j} = \mathbf{k}\times\mathbf{k} = 0$ with $\mathbf{i}\times\mathbf{j} = \mathbf{k}$, $\mathbf{j}\times\mathbf{k} = \mathbf{i}$, and $\mathbf{k}\times\mathbf{i} = \mathbf{j}$. In Figure 5.4, the vectors can be written in terms of components as $\mathbf{a} = a_x\mathbf{i} + a_y\mathbf{j}$, $\mathbf{b} = b_x\mathbf{i} + b_y\mathbf{j}$, and $\mathbf{c} = c_z\mathbf{k}$.

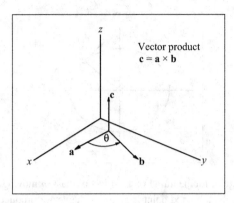

FIGURE 5.4   The vector product of vectors **a** and **b** is defined as **a** × **b** = **c** with resultant vector **c** oriented perpendicular to the *xy*-plane containing **a** and **b**. Vector **c** is directed according to the right-hand rule as described in the text. The tails of the vectors are made to coincide.

Using vector product notation, the angular momentum of an object of mass m undergoing circular motion with speed v and radius r is given by

$$\mathbf{l} = m\,\mathbf{r} \times \mathbf{v} = \mathbf{r} \times \mathbf{p} = (m\,r\,v\sin\theta)\hat{\mathbf{n}} = m\,r\,v\,\hat{\mathbf{n}} = m\,r^2\,\omega\hat{\mathbf{n}} \qquad (5.6)$$

In Equation (5.6), $\hat{\mathbf{n}}$ is a unit vector normal to the *rv*-plane with direction given by the right-hand rule. The angle $\theta = \pi/2$, since the vectors **r** and **v** are orthogonal as shown in Figure 5.5. Equation (5.6) can be written as $\mathbf{l} = m\,r^2\,\boldsymbol{\omega}$ with the angular momentum vector **l** and angular velocity vector $\boldsymbol{\omega}$ aligned parallel to the axis of rotation and perpendicular to the plane of motion. The tails of **l** and $\boldsymbol{\omega}$ coincide with the tail of **r**.

Equation (5.6) shows that the angular momentum l is proportional to the angular velocity ω since $v = \omega\,r$. In dealing with the rotational motion of objects, it is useful to introduce a quantity I called the moment of inertia. For a system consisting of a mass m at radial distance r from the axis of rotation $I = m\,r^2$. Like the mass of an object, the moment of inertia is a scalar quantity and is essential for discussing rigid body dynamics as will be shown in Chapter 6. For a system of several particles i with masses m_i at positions r_i with respect to some reference point, such as the selected origin, the moment of inertia is defined as $I = \sum_i m_i\,r_i^2$. For continuous systems, the summation is replaced by an integral over volume elements in the object of interest. As shown in Section 5.7, the moment of inertia plays a crucial role in modifying Newton's second law to apply to rotational dynamics.

With the introduction of I, Equation (5.6) for the angular momentum becomes $\mathbf{l} = m\,r^2\,\omega\hat{\mathbf{n}} = I\,\omega\hat{\mathbf{n}}$. Just as the linear velocity **v** is a vector, so too is the angular velocity $\boldsymbol{\omega} = \omega\hat{\mathbf{n}}$, which, like the angular momentum **l**, is directed along the axis of rotation. The vector product relationship $\mathbf{v} = \boldsymbol{\omega} \times \mathbf{r}$ links **v** and $\boldsymbol{\omega}$.

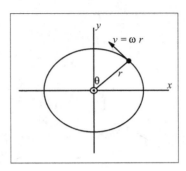

FIGURE 5.5   The angular momentum **l** of an object of mass m moving with constant speed v in a circular trajectory about a fixed point is given by the vector product $\mathbf{l} = \mathbf{r} \times \mathbf{p} = m\,\mathbf{r} \times \mathbf{v}$. If **r** and **p** are in the xy-plane, as illustrated, then the angular momentum vector is directed along z, which points upwards out of the page.

5.7   NEWTON'S SECOND LAW FOR ROTATIONAL MOTION

Newton's second law $\mathbf{F} = \dfrac{d\mathbf{p}}{dt} = m\,\mathbf{a}$ for linear motion of an object of mass m relates the rate of change of linear momentum $\mathbf{p} = m\,\mathbf{v}$ to an applied force **F**. A similar relationship holds for the angular momentum **l** with the second law expressed as $\Gamma = \dfrac{d\mathbf{l}}{dt} = I\dfrac{d\omega}{dt} = I\,\alpha$ where Γ is the torque producing the change in **l**, ω is the angular velocity, α is the angular acceleration, and I is the moment of inertia as defined in Section 5.6. Justification for this modified form of Newtons' second law, $\Gamma = I\,\alpha$, is given below.

Consider a simple system consisting of an object of mass m attached to a pivot by a very light rod, of length r and negligible mass. The object is caused to undergo circular motion about the pivot. In order to change the angular momentum of the rotating mass, it is necessary that a rotating force, called a torque, be applied to the mass itself or to a point on the rod or in some other way involving the pivot. Experience shows that a rotating force becomes most effective when, firstly, the point of application is moved outwards from the pivot towards the mass and, secondly, the force is kept perpendicular to the rod. This suggests that the torque vector be defined as $\Gamma = \mathbf{r} \times \mathbf{F} = (r\,F\sin\phi)\hat{\mathbf{n}}$ with ϕ the angle subtended by vectors **r** and **F**. The torque is a maximum for $\phi = \pi/2$ when **r** is perpendicular to **F**. The right-hand rule then shows that the torque vector is aligned along the rotation axis. As indicated above, a convenient way to apply such a torque for this simple mass-rod system is to attach the radial light rod to a vertical drive shaft, which can be caused to rotate by a drive mechanism. Ideally, the drive mechanism should be capable of being decoupled when necessary to allow free rotation of the mass and rod with constant angular velocity.

The considerations given above for a system consisting of a mass attached by a light rod to a pivot suggest that for rotational motion Newton's second law should be modified by replacing the force $\mathbf{F}$ by the torque $\Gamma = \mathbf{r} \times \mathbf{F}$, and the linear momentum $\mathbf{p}$ by the angular momentum $\mathbf{l} = \mathbf{r} \times \mathbf{p}$. This procedure gives Newton's second law for rotational motion as

$$\Gamma = \frac{d\mathbf{l}}{dt} = I \frac{d\omega}{dt} = I\alpha \qquad (5.7)$$

This generalization of Newton's second law is of fundamental importance in considering the dynamics of rigid bodies, as discussed in Chapter 6.

Note that if opposing torques of equal magnitude are applied to a body that can undergo rotational motion, then no motion will occur. The system remains in static equilibrium.

Exercise 5.4: An object of mass 2 kg is attached by a rod of negligible mass and length 50 cm to a central pivot. The mass is supported by a low-friction horizontal air table and is caused to undergo circular motion. If the mass makes 20 revolutions per minute, what is the central force exerted by the rod on the mass? If a torque of 10 N m is applied to the system via a vertical shaft aligned along the axis of rotation, how long will it take to double the angular velocity of the rotating mass?

Initially the time for one revolution is $T = 60/20 = 3$ s and the angular velocity is $\omega = 2\pi/T = 2\pi/3 = 2.09$ rad/s. The acceleration towards the centre of rotation is $a_c = \omega^2 r = 2.09^2 \times 0.5 = 2.18$ m/s^2. The central force exerted by the light rod on the rotating mass is $F = ma_c = 2 \times 2.18 = 4.36$ N.

The moment of inertia of the rotating mass-rod system about the rotation axis is $I = mr^2 = 2 \times 0.25 = 0.5$ kg m^2. The angular acceleration is $\alpha = \Gamma/I = 10/0.5 = 20$ rad/s^2. From the first rotational kinematic equation, $\omega = \omega_0 + \alpha t$ as given in Equation (5.1), it follows that the time taken for the angular velocity to double is $t = (\omega - \omega_0)/\alpha = 2.09/20 = 0.1$ s.

5.8   ROTATIONAL KINETIC ENERGY

By adapting the expression for the kinetic energy of a mass m undergoing linear motion, as given in Chapter 4, the kinetic energy of a mass m undergoing circular motion with angular velocity ω and radius r is written as

$$K = \frac{1}{2}mv^2 = \frac{1}{2}m\omega^2 r^2 = \frac{1}{2}I\omega^2 \qquad (5.8)$$

with $I = mr^2$ the moment of inertia about the rotation axis as given in Section 5.4. The angular velocity of the object can be increased by applying a torque that does work on the system. If friction effects are negligible, the work done by the applied torque in increasing the angular velocity from ω_i to ω_f is equal to the change in the kinetic energy, giving

$$\Delta K = W = \frac{1}{2} I \left(\omega_f^2 - \omega_i^2 \right) \tag{5.9}$$

It is now necessary to obtain an expression for W in order to connect the applied torque to the corresponding change in K. In the case of a linear displacement, the work done by an applied force F in moving its point of application through a distance δx parallel to the force is given by $F \, \delta x$. The expression for the work done by a torque involves the angular displacement it produces. Consider the simple rotating system consisting of a mass undergoing circular motion, as dealt with previously in Section 5.7. The work δW done by torque Γ in producing an angular displacement $\delta \theta$ of the mass is $\delta W = \Gamma \, \delta \theta$. This result is obtained as follows. Consider a force F acting on the mass at some point in its circular orbit. The tangential component of F in the plane of motion, designated $F_\perp$ since it acts perpendicular to the radius r, does work $\delta W = F_\perp \, r \, \delta \theta = (\mathbf{r} \times \mathbf{F}) \cdot \hat{\mathbf{n}} \, \delta \theta = (r \, F \sin \phi) \, \delta \theta = \Gamma \, \delta \theta$ where, as an approximation, it is assumed that for small $\delta \theta$ the orbital displacement $r \, \delta \theta$ is approximately linear. The angle $\phi = \pi/2$ since r and F are perpendicular to one another. The total work W done by a constant torque Γ in a large angular displacement is obtained by summing the elementary contributions δW. Converting the sum to an integral gives the required result,

$$W = \int_{\theta_i}^{\theta_f} \Gamma \, d\theta = \Gamma \left(\theta_f - \theta_i \right) = \Gamma \, \Delta \theta \tag{5.10}$$

Taken together, the expressions for W given in Equations (5.9) and (5.10) lead to the following relationship:

$$W = \Gamma \, \Delta \theta = \frac{1}{2} I \left(\omega_f^2 - \omega_i^2 \right) \tag{5.11}$$

For a mass undergoing circular motion, Equation (5.11) relates the work done by an applied torque to the change in kinetic energy ΔK expressed in terms of the moment of inertia and the change in the square of the angular velocity.

Exercise 5.5: A mass of 8 kg, which is attached by a light rod of length 60 cm to a central pivot, executes circular motion on a horizontal low friction air table at 45 revolutions per minute (rpm). If a braking torque of 0.4 N m is applied to stop the motion, how many revolutions will it take for the mass to come to rest?

From Equation (5.11), the angular displacement which occurs during the braking period is given by $\Delta \theta = \dfrac{I \left(\omega_i^2 \right)}{2\Gamma}$. Note that a sign change has been made because the applied torque slows the rotational motion. The moment of inertia is $I = mr^2 = 8 \times 0.6^2 = 2.88 \, \text{kg m}^2$, and the initial angular velocity is $\omega_i = \dfrac{(45 \times 2\pi)}{60} = 4.71 \, \text{rad/s}$. Substitution of these values for I and ω_i into the previous expression for the angular displacement gives

$$\Delta\theta = \frac{2.88 \times 7.34}{2 \times 0.4} = 26.4 \, \text{rad}, \text{ or } 4.2 \text{ revolutions.}$$

Many of the results obtained for a small object of mass m undergoing circular motion can be generalized to the rotation of large rigid bodies which are regarded as collections of elementary masses. The concepts of angular displacement, angular velocity, angular acceleration, torque, and moment of inertia are readily extended to the pivoted motion of large objects. Chapter 6 deals with rigid body dynamics.

5.9 PLANETARY MOTION AND KEPLER'S LAWS

The Earth orbits the Sun along a path that is close to circular but is actually slightly elliptical. The law of universal gravitation was used by Newton to explain the orbital motion of the Earth and the other planets about the Sun in mathematical terms following the establishment of details of the trajectories by Johannes Kepler in the seventeenth century. Kepler's analysis of Tycho Brahe's careful astronomical observations led him to summarize his findings on planetary orbits in three famous laws which are stated below.

- *Kepler's first law:* The planets follow elliptic orbits about the Sun with the Sun at a focal point.
- *Kepler's second law:* A line from the Sun to a planet sweeps out equal areas in equal times.
- *Kepler's third law:* The square of a planet's orbital period is proportional to the cube of the semi-major axis of its elliptic path.

Figure 5.6 shows a representative elliptic orbit with two focal points, labelled F_1 and F_2, and the Sun located at F_1.

The form of an elliptic orbit is determined by a quantity called the eccentricity, which is defined as $\varepsilon = \sqrt{a^2 - b^2}/a$ with a the length of the semi-major axis and b that of the semi-minor axis. If $b = a$, then $\varepsilon = 0$ and the orbit is circular. Otherwise, elliptic orbits have $b < a$, and $\varepsilon > 0$ with a maximum value of unity. The orbit in Figure 5.6 has $\varepsilon = 0.63$. Only two of the eight planets in the solar system, Mercury with $\varepsilon = 0.2$ and Mars with $\varepsilon = 0.09$, have eccentricities greater than 0.06. Venus, $\varepsilon = 0.01$, and Earth, $\varepsilon = 0.02$, both have very low eccentricities. Interestingly, the almost circular orbits of the planets lie roughly in the same plane called the ecliptic plane. This feature suggests that the planets were formed from a disk of material that originally encircled the Sun.

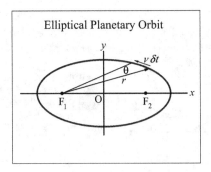

FIGURE 5.6 A representative elliptical planetary orbit with the Sun located at the focal point F_1. Apart from the planet Mercury, the orbits of the planets in the solar system are close to circular and only slightly elliptical.

Kepler's first and third laws are a direct result of the inverse square law of gravitation. With regard to the first law, it can be shown that there are four different trajectories which can arise when an object of mass m is in motion in the gravitational field produced by a very large mass M. These possible trajectories are circular, elliptic, parabolic, and hyperbolic. Only the circular and elliptic orbits are closed. Parabolic and hyperbolic trajectories correspond to the deflection of a high energy object which then heads off into deep space.

For circular orbits, it is straightforward to show that Kepler's third law can be explained by using Newton's second law of motion together with the law of universal gravitation. Although it is convenient to assume that the planet's orbit is circular in order to simplify the analysis, Kepler's third law applies quite generally to all closed orbits. From Equation (5.4), the centripetal acceleration for a circular motion

of radius r is $a = -\omega^2 r = \left(\dfrac{2\pi}{T}\right)^2 r$ where T is the orbital period. Newton's second

law gives $\dfrac{G M m}{r^2} = m\left(\dfrac{2\pi}{T}\right)^2 r$ and it follows that

$$T^2 = \frac{4\pi^2}{G M} r^3 \qquad (5.12)$$

This is the required result to justify Kepler's third law.

The area of the triangle that is swept out in a time δt by a line of length r connecting an orbiting planet to the Sun at focal point F_1, as shown in Figure 5.6, is given by $\delta A = \dfrac{1}{2}(r\sin\theta)v\,\delta t$. Since the instantaneous linear momentum is $p = m v$, the rate at which area is swept out can be written as follows:

$$\frac{\delta A}{\delta t} = \frac{1}{2}\frac{r\,p\sin\theta}{m} = \frac{1}{2m}\left|\mathbf{r}\times\mathbf{p}\right|$$

Using the expression for the angular momentum l given in Equation (5.6) leads to the relationship

$$\frac{\delta A}{\delta t} = \frac{l}{2m} \qquad (5.13)$$

Since the gravitational force acts along r towards the Sun, there is no torque to change the angular momentum of the planet about the Sun. The right-hand side of Equation (5.13) is therefore constant, showing that $\frac{\delta A}{\delta t}$ is constant as required by Kepler's second law. Newton's explanation of Kepler's laws of planetary motion using the gravitational inverse square law also applies to the orbital motion of both satellites and the Moon about the Earth. Newton could offer no explanation for the underlying mechanism, which gives rise to this inverse square law involving interaction at a distance between objects with mass. The law is introduced as a requirement to account for experimental observations. In the early twentieth century, Albert Einstein put forward his general relativity theory, which provides a deeper insight into gravitational phenomena. For example, general relativity predicts the existence of gravitational waves, which travel at the speed of light and can arise from the interaction of massive bodies including neutron stars or black holes. These waves were first observed in 2015 using extremely sensitive detectors. The detected waves had travelled a distance of 1.4 billion light years to reach the Earth and were attributed to the merging of two black holes with masses of around 30 and 35 solar masses. Gravitational wave detection has opened up a new field of astronomy. The law of universal gravitation used by Newton is obtained as a prediction of Einstein's general relativity theory for applications involving interactions between well-separated massive objects such as those in the solar system. The amplitudes of the gravitational waves emitted by these interactions are much too small to be detected by the most sensitive detectors that have been constructed.

5.10 SATELLITE ORBITS

The gravitational interaction plays a central role in describing the structure of the universe. Each galaxy has a massive black hole at its centre, around which stars orbit. For example, the Sun in our Milky Way galaxy is 25×10^3 light years from the galaxy centre and takes 250×10^6 years to complete an orbit. On a more modest scale, the motion of the Sun's planets, including the Earth plus its satellites, are well described by Newton's law of universal gravitation. Important examples of the Earth's satellites are the International Space Station, the Hubble Space Telescope, and navigation and weather satellites. The James Webb infrared telescope, which released its first images in 2022, has opened up new opportunities for deep space astronomy. The trajectories of Earth-orbiting satellites are in many cases close to circular and have altitudes h above the Earth's surface ranging from low-Earth orbit $(h < 2000 \text{ km})$, through medium-Earth orbit $(2000 < h < 36,000 \text{ km})$, to high-Earth orbit $(h > 36,000 \text{ km})$. Illustrative examples are given below. While satellite orbits may not be circular, and some are markedly elliptical, it will, for simplicity, be assumed that the orbits are to a good approximation circular.

Exercise 5.6: A geosynchronous weather satellite orbits the Earth once per day keeping its position directly above a chosen place on the Earth's surface. What is the radius R of the satellites circular orbit? Determine the angular velocity of the satellite. Take $G = 6.67 \times 10^{-11}$ N m^2/kg^2 and $M_E = 5.97 \times 10^{24}$ kg.

The universal gravitation law, together with Newton's second law for a mass m undergoing circular motion about the Earth, leads to the expression given in Equation (5.12), which here takes the form $T^2 = \dfrac{4\pi^2}{G M_E} R^3$, where T is the orbital period and M_E is the Earth's mass. Rearranging gives

$$R^3 = \frac{G M_E}{4\pi^2} T^2 = \left(\frac{6.67 \times 10^{-11} \times 5.97 \times 10^{24}}{39.48} \right) \times (24 \times 3600)^2 = 7.5 \times 10^{22} \text{ m}^3$$

The radius measured to the centre of the Earth is thus $R = 4.22 \times 10^4$ km. Allowing for the Earth's radius $R_E = 6370$ km gives the geosynchronous altitude above the Earth's surface as roughly 36×10^3 km. The angular velocity of the satellite is $\omega = v/R = \left(\dfrac{2\pi R}{T} \right)/R = \dfrac{2\pi}{24 \times 3600} = 7.27 \times 10^{-5}$ rad/s.

Exercise 5.7: The International Space Station is in a low Earth orbit with a radius that can be approximated by the Earth's radius. Determine the orbital period of the Space Station.

Inserting numbers into Kepler's third law expression, $T^2 = \dfrac{4\pi^2}{G M_E} R^3$, gives

$$T^2 = \left(\frac{39.48}{6.67 \times 10^{-11} \times 5.97 \times 10^{24}} \right) \times \left(6.37 \times 10^6 \right)^3 = 2.56 \times 10^7 \text{ s}^2$$

The Space Station's orbital period is $T = 5060$ s $= 84$ minutes.

Note that for all the Earth's satellites, the quantity $\dfrac{4\pi^2}{G M_E} = 9.90 \times 10^{-14}$ s^2/m^3 is a constant whose value is determined by the mass of the Earth, M_E, and the gravitational constant, G.

Most of the Earth's satellites are in low-Earth orbit, while weather satellites and global position satellites are in medium-Earth orbit. The Moon, which is a natural high-Earth orbit satellite, with an average orbit radius $R_M = 3.85 \times 10^5$ km, has a period T_M = 28 days. It should be noted that the Earth and a satellite orbit about their common centre of mass, which for comparatively low-mass artificial satellites is very close to the centre of mass of the Earth. This closeness no longer holds for the orbiting Moon, which has a mass roughly one-hundredth that of the Earth. For the Earth-Moon system, the centre of mass is about 4670 km (~ 0.73 R_E) from the Earth's centre. The centre of mass concept for extended objects is discussed in Chapter 6.

6 Rigid Body Motion

6.1 INTRODUCTION

The development of a physical understanding of the rotational motion of extended objects such as wheels, rollers, and gears, is of considerable practical importance. While it might seem that the variety of shapes and sizes of rigid bodies would lead to complexity in analysing their rotational dynamics, it is possible to give a generalized description in terms of the unifying concepts that are introduced in Chapter 5 for dealing with the circular motion of a localized mass subject to a central force. For example, the concepts of angular displacement and angular velocity of the many elementary pieces that make up a rigid body play a central role in the description of the rotational motion of such bodies.

In dealing with rigid body motion, it is instructive, as a starting point, to consider the conditions for these objects to be stationary in a particular frame of reference as discussed in Section 6.2. It is then necessary to define both the centre of mass and the moment of inertia for these bodies. Generalization of Newton's second law to the translational and rotational motion of extended objects, together with the concepts of work and kinetic energy for rotating systems, provides the basis for dealing with a large variety of situations, including the spinning of disks and the rolling motion of cylinders.

6.2 EQUILIBRIUM CONDITIONS

Consider a rigid body that is acted on by a combination of external forces $\mathbf{F}_i$ and torques $\mathbf{\Gamma}_i$. If the object is at rest in a chosen frame of reference, the following two static equilibrium conditions must hold in that frame: $\sum_i \mathbf{F}_i = \mathbf{0}$ and $\sum_i \mathbf{\Gamma}_i = \mathbf{0}$, where

the summations signify the vector sums of forces and torques, respectively. While internal forces, due to interactions between the constituent atoms, are present at the microscopic level in the material of the body, these forces obey Newton's third law and cancel in pairs. Internal forces therefore play no role in the macroscopic translational or rotational motion of a rigid body. If one, or both, of the equilibrium conditions are not met, the body will execute translational and/or rotational motion as given by Newton's second law.

DOI: 10.1201/9781003485537-6

As a simple illustrative example of static equilibrium, consider a uniform rod which is stationary on a horizontal surface situated near the Earth's surface, with the downward gravitational weight force matched by an equal and opposite upward reaction force from the tabletop. If a horizontal force **F** is applied at the midpoint of the rod, it will start to move unless an equal and opposite force is applied to match the force **F**. Alternatively, two opposing forces of magnitude **F**/2 could be applied near the two ends of the rod, which would also balance both the torques and the forces on the rod. Many other equilibrium arrangements could be considered, as discussed in detail in Section 6.7.

6.3 CENTRE OF MASS AND CENTRE OF GRAVITY

Tracking the motion of the centres of mass of rigid bodies in motion is important in developing an understanding of the dynamics of extended objects. For example, when considering the motion of a satellite in orbit around the Earth, it is necessary to grasp that the motion of the two bodies occurs about their common centre of mass. Because the Earth's mass is so much larger than that of a satellite, the centre of mass is close to the centre of the Earth, but not exactly at its centre. This situation is discussed in Section 6.3.1. A related concept to that of centre of mass is that of centre of gravity, which involves the torques on a rigid body situated, for example, in the Earth's gravitational field. If the gravitational field is uniform, then the centre of mass and the centre of gravity coincide.

6.3.1 CENTRE OF MASS

In introducing the centre of mass, it is convenient to consider firstly a set of discrete masses m_i, where $i = 1, 2, 3, \ldots$, with position vectors $\mathbf{r}_i$ in 3D space. Denoting the position of the centre of mass in a chosen coordinate system by $\mathbf{r}_{CM}$, and requiring that the discrete masses be distributed about the centre of mass according to the condition $\sum_i m_i (\mathbf{r}_i - \mathbf{r}_{CM}) = 0$, gives $\sum_i m_i \, \mathbf{r}_i - \mathbf{r}_{CM} \sum m_i = 0$, and hence

$$\mathbf{r}_{CM} = \frac{1}{M} \sum_i m_i \, \mathbf{r}_i \qquad (6.1)$$

where $M = \sum_i m_i$ is the total mass of the system.

Consider two masses m_1 and m_2, with $m_2 > m_1$ and their centres separated by a distance L as illustrated in Figure 6.1. The vector **r** joins the centre of mass 1 to the centre of mass 2.

By symmetry, the centre of mass lies on the direction of **r**. The origin is chosen to be at the midpoint, a distance $L/2$ from each mass, with the centre of mass at r_{CM}. The use of Equation (6.1) gives the position of the centre of mass as

$$r_{CM} = \frac{\left(-\frac{1}{2} m_1 L + \frac{1}{2} m_2 L\right)}{\left(m_1 + m_2\right)} = \frac{\left(m_2 - m_1\right) L}{\left(m_1 + m_2\right) 2} \qquad (6.2)$$

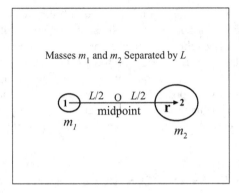

FIGURE 6.1   The position of the centre of mass for the two masses m_1 and m_2 whose centres are separated by a distance L is given by Equation (6.2).

If $m_1 = m_2$, then $r_{CM} = 0$ and the centre of mass is at the origin, midway between the two masses. For $m_2 \gg m_1$, Equation (6.2) shows that $r_{CM} \approx L/2$ which is close to, or even inside, the large mass m_2 as expected. In general, it follows that the centre of mass lies in the range given by $0 \le r_{CM} \le L/2$. Note that the choice of origin position is arbitrary, and equivalent expressions to those given above for r_{CM} can be obtained using other origin locations.

For continuous mass distributions, it is necessary to replace the summation in Equation (6.1) by an integral over the volume of the rigid body involved. Equation (6.1) becomes

$$\mathbf{r}_{CM} = \frac{1}{M} \int_V \rho(\mathbf{r})\, \mathbf{r}\, dV \tag{6.3}$$

where $\rho(\mathbf{r})\, dV$ is a mass element at $\mathbf{r}$ with volume dV and local mass density $\rho(\mathbf{r})$. For homogeneous materials, $\rho(\mathbf{r}) = \rho = M/V$ is a constant characteristic of the material. Equation (6.3) then takes the form

$$\mathbf{r}_{CM} = \frac{1}{V} \int_V \mathbf{r}\, dV \tag{6.4}$$

In 3D Cartesian coordinates, $dV = dx\, dy\, dz$, and the integral becomes a triple integral over x, y, and z. The position of the centre of mass is given by $\mathbf{r}_{CM} = x_{CM}\, \mathbf{i} + y_{CM}\, \mathbf{j} + z_{CM}\, \mathbf{k}$. Use of Cartesian coordinates in Equation (6.4) leads to the integral for x_{CM} as

$$x_{CM} = \frac{1}{V} \iiint x\, dx\, dy\, dz \tag{6.5}$$

Similar integrals are obtained for y_{CM} and z_{CM}. The upper and lower limits in each integral are connected to the corresponding dimensions of the body.

Exercise 6.1: Determine the position of the centre of mass of a solid cube of edge length L using Cartesian coordinates to specify the shape.

The volume of the cube is $V = L^3$. Equation (6.5), with $\int_0^L\int_0^L dy\, dz = L^2$,

gives $x_{CM} = \dfrac{L^2}{V}\int_0^L x\, dx = \dfrac{L^2}{L^3} \times \left(\dfrac{L^2}{2}\right) = \dfrac{L}{2}$.

This result shows that the centre of mass is at the midpoint along x. In similar fashion $y_{CM} = z_{CM} = L/2$ and it follows that the centre of mass is located at the geometrical centre of the cube.

For symmetrical objects such as spheres, cylinders, and parallelepipeds, it is possible to use symmetry properties to determine the position of the centre of mass. For example, the centre of mass of a sphere is at the geometrical centre of the sphere.

6.3.2 Centre of Gravity

As the name implies, the centre of gravity concept is concerned with a system of masses on which gravitational forces act. The total downward force acts through the centre of gravity of the system, as discussed below. In effect, the total mass is located at the centre of gravity. In equilibrium, the upward reaction force, provided by a support of some type, acts to match the downward gravitational force. It is necessary to allow for clockwise and counterclockwise torques about the centre of gravity, and in equilibrium these must also sum to zero. Note that if the gravitational field is uniform over the system of masses, then the centre of mass and the centre of gravity coincide. This result is readily obtained for a model system of two masses, and follows quite generally for larger systems, which are viewed as collections of pairs of masses.

Consider again the two masses m_1 and m_2, with $m_2 > m_1$, connected by a very light but rigid rod of length R and suspended on a movable support in the Earth's gravitational field as illustrated in Figure 6.2. It is convenient to choose the x-axis to lie along the line joining the two masses with the origin located at the midpoint. Mass m_1 is at $-R/2$ and mass m_2 is at $R/2$. In equilibrium, the reaction force $F_R = (m_1 + m_2)g$ acts upwards through the pivot located at r_{CG}. Balancing clockwise and counterclockwise torques gives $-\dfrac{1}{2}m_1\, g\, R + \dfrac{1}{2}m_2\, g\, R - (m_1 + m_2)g\, r_{CG} = 0$ and it follows that

$$r_{CG} = \frac{(m_2 - m_1)}{(m_1 + m_2)}\frac{R}{2} \tag{6.6}$$

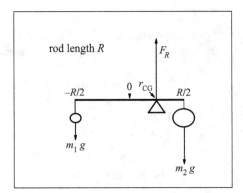

FIGURE 6.2 Equilibrium of two masses m_1 and m_2 connected by a light rod of length R which rests on a support. The support is located at the centre of gravity, a distance r_{CM} from the rod midpoint at 0.

Note that for $m_2 \gg m_1$, the centre of gravity is found at $r_{CG} \approx R/2$, very close to m_2. The expressions obtained for r_{CG} in Equation (6.6) and for r_{CM} in Equation (6.2) are identical, showing that the centre of mass and the centre of gravity coincide as expected since the Earth's gravitational field is locally uniform.

The selected origin can be situated at any point in the vicinity of the two masses when determining the position of the centre of gravity. As an illustration of this point, the origin is shifted to coincide with mass m_1 at $-R/2$. In this case, the sum of torques expression becomes $m_2\, R\, g - \left(m_1 + m_2 \right) r_{CG}\, g = 0$, and this gives $r_{CG} = \dfrac{m_2}{\left(m_1 + m_2 \right)} R$. For $m_1 = m_2$ the centre of gravity is at $R/2$, while in the limit $m_2 \gg m_1$ the centre of gravity is close to m_2 with $r_{CG} \approx R$.

In order to determine the centre of gravity of an arbitrarily shaped rigid body, it is necessary to consider dimensions higher than 1D and to integrate over a distribution of volume elements each of which contributes a mass $dm = \rho\, dV$ to the total mass of the object. Assuming that the density ρ is constant over the volume of the object, the procedure is essentially similar to that used above for discrete masses. The location of a given volume element is specified by its position vector $\mathbf{r}$ referred to the origin of a 3D set of coordinates, with the centre of gravity given by

$$\mathbf{r}_{CG} = \frac{\rho}{M} \int_V \mathbf{r}\, dV \qquad (6.7)$$

Equation (6.7) is equivalent to Equation (6.4) for the centre of mass since the gravitational field is assumed to be uniform over the volume of the rigid body and the two centres therefore coincide.

6.4 TRANSLATIONAL AND ROTATIONAL MOTION

The equations governing the linear translational dynamics of a rigid body are essentially the same as those for a single particle or a small object of mass m. As pointed out in Section 6.3, a solid object is viewed as a collection of elements of volume dV and mass $\rho\,dV$. The density ρ is assumed constant over the volume of a homogeneous material. The total mass M is given by the integral $M = \int_V \rho\,dV$ carried out over the volume of the object. For a body in linear motion, all the volume elements have the same velocity $\mathbf{v}$, and the total momentum is given by $\mathbf{p} = M\,\mathbf{v}$. Newton's second law is expressed as $\mathbf{F} = \dfrac{d\mathbf{p}}{dt} = M\,\mathbf{a}$, where $\mathbf{F}$ is the force acting on a body and $\mathbf{a}$ is the body's acceleration. In collision processes involving the transfer of linear momentum between objects, the objects behave as point masses located at their centres of mass.

6.4.1 MOMENT OF INERTIA

The rotational motion of a rigid body involves several of the concepts introduced in Chapter 5 when describing the circular motion of a mass m subject to a central force. This correspondence of basic concepts can be understood by considering the motion of small volume elements in a rigid body, which is rotating about a fixed axis. All the volume elements, labelled i, execute a circular motion with radius r_i given by the distance of the element from the axis of rotation. The angular displacement $\Delta\theta$ is the same for all the elements. Similarly, the angular velocity $\omega = \dfrac{d\theta}{dt}$ and the angular acceleration $\alpha = \dfrac{d\omega}{dt}$ have common values for all the elements. The concepts of moment of inertia $I = m\,r^2$, angular momentum $L = m\,v\,r = I\,\omega$, and kinetic energy of rotation $K = \dfrac{1}{2}I\,\omega^2$, introduced in Chapter 5 for a single particle undergoing circular motion with radius r, are readily generalized to apply to the rotational motion of a solid body. This is done by extending the moment of inertia definition to an assembly of volume elements dV_i. For a rigid body of uniform density ρ the moment of inertia is thus defined as $I = \sum_i m_i\,r_i^2 = \sum_i \rho\,r_i^2\,dV_i$. Converting the sum to an integral over the volume of the rigid body gives

$$I = \rho\int_V r^2\,dV \qquad (6.8)$$

Expressions for the moments of inertia of rigid bodies with various shapes are calculated below.

Exercise 6.2: (a) Determine the moment of inertia of a solid cylinder of density ρ, radius R, and length L about the long central axis as depicted in Figure 6.3. (b) Determine the moment of inertia for a narrow cylindrical rod, with a diameter much less than its length, about an axis through the centre perpendicular to the long axis.

(a) It is convenient to consider the cylinder to be made up of a stack of thin disks each of thickness ΔL. A representative disk of area $A = \pi R^2$ and thickness ΔL, centred at point O, which is also the centre of mass of the cylinder, is shown in Figure 6.3. Symmetry considerations suggest that volume elements dV should be chosen in the form of concentric rings of radius r, and width dr. Substituting $dV = 2\pi r\, dr\, \Delta L$ in Equation (6.8)

gives $I_{\text{disk}} = \rho \int_0^R (2\pi r\, \Delta L) r^2\, dr = 2\pi \rho\, \Delta L \int_0^R r^3 dr = \frac{1}{2}\rho A\, \Delta L\, R^2 = \frac{1}{2}\Delta M\, R^2$.

The moment of inertia of the cylinder is then obtained by summing over the stack of disks. Converting the sum to an integral gives the required

result $I_{\text{cylinder}} = \frac{1}{2}\rho A R^2 \int_{-L/2}^{L/2} dL = \frac{1}{2}\rho A L R^2 = \frac{1}{2}M R^2$. Note that this form

holds for both thin disks and cylinders.

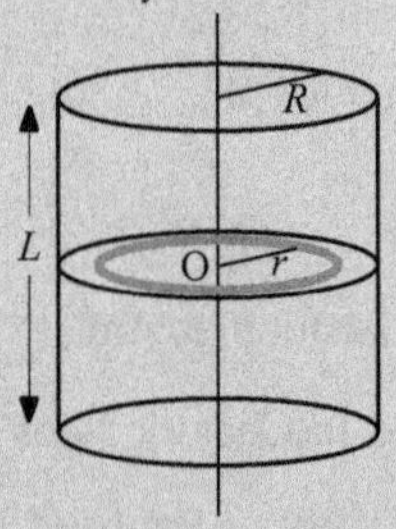

Cylinder axis

FIGURE 6.3 The solid cylinder shown has mass M, length L, and radius R. The moment of inertia is determined about the cylinder axis by integrating over a stack of disk-shaped elements. The shaded ring represents a volume element in a representative disk. The centre of mass of the cylinder is at the origin O.

(b) In this case, which is effectively 1D, use is again made of thin disk-shaped volume elements. A sketch of the rod with a representative volume element is shown in Figure. 6.4.

A representative disk has thickness dl, diameter R, and is positioned at a distance l, in the range from $-L/2$ to $+L/2$, along the axis through the centre

of the rod. Equation (6.8) leads to the result $I = 2\pi \rho R^2 \int_0^{L/2} l^2\, dl = \frac{1}{12}M L^2$.

The same result would hold as a good approximation for a rectangular long thin rod with cross-sectional dimensions much less than its length.

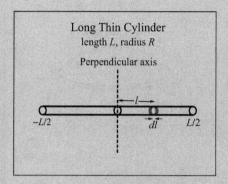

FIGURE 6.4 The long rod shown has length L, and radius R much less than the length. The moment of inertia about an axis through the centre perpendicular to the rod is obtained by summing contributions from disk-shaped elements of length dl.

In the general case, in which the width of the rod is not much less than its length, a different expression is obtained using the parallel axis theorem, which is introduced below.

Exercise 6.3: Determine the moment of inertia of a solid sphere of radius R and density ρ about an axis through the centre. The sphere is depicted in Figure 6.5.

The centre of mass of the uniform sphere is located at its centre, labelled O in Figure 6.5, with the rotation axis along z. It is convenient to view the sphere as a stack of thin circular disks oriented perpendicular to the z-axis, with the radius r of a particular disk, which has its centre at a distance z from O given by $r = \sqrt{R^2 - z^2}$. Exercise 6.2 (a) gives the moment of inertia of a disk of mass m, density ρ, thickness dz, and radius r about an axis through the centre as $I = \dfrac{1}{2} m\, r^2$ with $m = \rho\, \pi \left(R^2 - z^2\right) dz$. The moment of inertia of the sphere about z is obtained by summing the up contributions of the stacked disks, of varying radii, which make up the total volume. Converting the sum to an integral over z leads to the relationship $I = 2 \times \dfrac{1}{2} \rho\, \pi \displaystyle\int_0^R \left(R^2 - z^2\right)^2 dz = \rho\, \pi \displaystyle\int_0^R \left(R^4 - 2R^2 z^2 + z^4\right) dz =$

$\rho\, \pi R^5 \left(1 - \dfrac{2}{3} + \dfrac{1}{5}\right) = \dfrac{2}{5} M\, R^2$, with $M = \dfrac{4}{3}\rho\, \pi R^3$. The required result is

$I_{\text{sphere}} = \dfrac{2}{5} M\, R^2$.

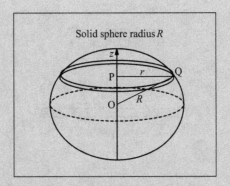

FIGURE 6.5   The solid sphere of radius R is viewed as a stack of disks of radii r which fit inside the sphere. A representative disk is shown with radius $r = \sqrt{R^2 - z^2}$ and thickness dz.

6.4.2 THE PARALLEL AXIS THEOREM

The parallel axis theorem simplifies the calculation of the moment of inertia I of a solid object about an axis that does not pass through its centre of mass, by expressing I in terms of the moment of inertia I_{CM} about a parallel axis through the centre of mass. The proof of the theorem is straightforward.

Consider an object of mass M, volume V, and uniform density ρ with an attached set of Cartesian axes. The origin O is located at the centre of mass of the object, with the z-axis chosen parallel to the axis z' about which the moment of inertia is required, as shown in Figure 6.6. The z' axis intersects the xy-plane at point O$'$ with coordinates $x = a$ and $y = b$.

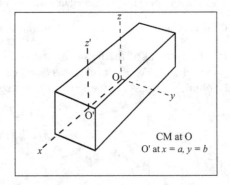

FIGURE 6.6   A solid object of mass M is depicted with centre of mass at the origin O in a Cartesian frame. The parallel axis theorem relates the moment of inertia about the z–axis, which passes through O, to the moment of inertia about axis z', which passes through O$'$ in the xy-plane at $x = a$ and $y = b$.

Using Equation (6.8), the moment of inertia about the z-axis involves an integral over the volume elements dV making up the object. The volume integral for I is written as a triple integral, involving the three Cartesian coordinates, in the form $I_{CM} = \rho \int_V r^2 dV = \rho \iiint_V (x^2 + y^2) dx\, dy\, dz$ where $dV = dx\, dy\, dz$ is a volume element at a distance r from the z-axis, with coordinates (x, y, z), giving the square of the distance as $r^2 = x^2 + y^2$.

The moment of inertia about the z'-axis is similarly given by $I_{\parallel} = \rho \iiint_v \left[(x-a)^2 + (y-b)^2 \right] dx\, dy\, dz$. Expanding the squared terms and grouping gives $I_{\parallel} = \rho \iiint_v \left[(x^2 + y^2) + (a^2 + b^2) - (2a\, x + 2b\, y) \right] dx\, dy\, dz$. The triple integrals $\rho\, 2a \iiint_v x\, dx\, dy\, dz$ and $\rho\, 2b \iiint_v y\, dx\, dy\, dz$, which involve linear terms in x and y, give, apart from multiplying constants, the coordinates of the centre of mass, as can be seen from Equation (6.5). Both these integrals therefore vanish, because the centre of mass is at the origin O. With $h^2 = a^2 + b^2$, the moment of inertia about z' becomes

$$I_{\parallel} = I_{CM} + M\, h^2 \tag{6.9}$$

The theorem states that the moment of inertia $I_{\parallel}$ of an object of mass M about an axis located at a distance h from a parallel axis through the centre of mass is equal to the moment of inertia I_{CM} about the centre of mass axis plus a quantity given by $M\, h^2$.

Exercise 6.4: Use the expression $I_{CM} = \dfrac{1}{12} M L^2$ for the moment of inertia of a solid narrow cylinder of length L and mass M about a perpendicular axis through the centre of mass, as determined in Exercise 6.2(b), to obtain the moment of inertia I about a parallel axis passing through one end of the cylinder.

The parallel axis theorem gives $I = \dfrac{1}{12} M L^2 + \dfrac{1}{4} M L^2 = \dfrac{1}{3} M L^2$.

6.5 ROTATIONAL DYNAMICS

The basic concepts and relationships that are needed for analysing the rotational motion of a rigid body about a chosen axis can be summarized in three points as follows:

1. Newton's second law for rotational motion is of central importance and takes the form $\Gamma = I\alpha$, where Γ is the torque producing rotation, I the moment of inertia about the rotation axis, and α the angular acceleration.
2. If a torque Γ is constant, it follows that α is constant and the rotational kinematic equations apply.

3. Accompanying any change in the angular velocity ω, there is a change in the angular momentum $\mathbf{L} = I\,\omega$.

The work–energy relationship is given by $\Gamma\,\Delta\theta = \Delta E$, with $\Delta E = \dfrac{1}{2}I\left(\omega_f^2 - \omega_i^2\right)$ provided friction is negligible and there is no change in the potential energy accompanying the rotational motion, a condition which holds when the rotation axis passes through the centre of gravity of the rotating object. The moment of inertia I about the rotation axis plays an important role in determining the dynamics.

Exercise 6.5: Flywheel-based energy storage and release systems have been developed to improve the efficiency of vehicles. Consider a flywheel of this type, made of carbon fibre material with a mass of 6 kg and a diameter of 20 cm, suspended in vacuum to permit angular velocities of 60,000 revolutions per minute (rpm) to be achieved. Energy is stored in the flywheel during braking periods and released via a drive mechanism when required. Calculate the moment of inertia of the flywheel about an axis through the centre of mass and perpendicular to the disk. Neglect energy contributions from the flywheel drive shaft. Determine the maximum energy that can be stored in the flywheel. If the angular velocity increases from 40,000 to 50,000 rpm over a braking time of 30 s, what is the average torque on the flywheel drive shaft?

The moment of inertia of the flywheel is $I = \dfrac{1}{2}M\,R^2 = 0.5\times 6\times 0.01 = 0.03$ kg m^2. At 60,000 rpm, the maximum stored energy is $E = \dfrac{1}{2}I\,\omega^2 = 0.5\times 0.03\times\left(2\pi\times 10^3\right)^2 = 5.92\times 10^5 = 164$ W-h.

The brake torque is given by $\Gamma = I\,\alpha$ with α the deceleration of the vehicle. From the kinematic equations for rotational motion, $\alpha = \dfrac{\delta\omega}{t} = \dfrac{2\pi\times 10^4}{60}\times\dfrac{1}{30} = 34.9$ rad/s^2. The torque on the flywheel drive shaft is therefore $\Gamma = I\,\alpha = 0.03\times 34.9 = 1.05$ N m.

Exercise 6.6: Consider a rotating oval rigid body for which the axis of rotation does not pass through the centre of gravity. Obtain an expression for the variation in the potential energy U as a function of angular displacement after the body is set into steady rotational motion in the Earth's gravitational field. Assuming that frictional effects can be neglected, use the law of mechanical energy conservation to determine the variation in the kinetic energy K with angular displacement.

The variation in the potential energy U of the body with orientation is obtained by considering the total mass of the object to be effectively located at the centre of mass, which coincides with the centre of gravity as discussed in

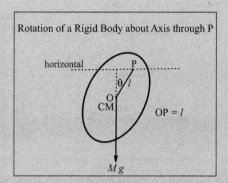

FIGURE 6.7 An oval-shaped rigid plate rotates clockwise about an axis through P with the centre of mass at O, a distance l from P. The angle θ specifies the orientation of OP with respect to the vertical direction. Friction effects are assumed to be negligible.

Section 6.3. Ignoring frictional effects, and using the law of mechanical energy conservation, it follows that the change in kinetic energy ΔK will be equal and opposite to the change in potential energy ΔU in order to keep the rotational energy constant.

If the centre of mass O is situated at a distance l from the rotation axis at P, then O rotates around P. Both ΔK and ΔU are functions of the angle θ shown in Figure 6.7. In considering changes in the potential energy, it is convenient to choose the reference level where $U = 0$ to correspond to $\theta = \pi/2$ with OP horizontal. It follows that $\Delta U = -M\,g\,l\cos\theta$ and therefore $\Delta K = \frac{1}{2}I\left(\omega^2 - \omega_0^2\right) = M\,g\,l\cos\theta$. The clockwise angular velocity ω has a maximum value when the centre of gravity is at its lowest point, with $\theta = 0$, and a minimum value when the centre of gravity is at its highest point, where $\theta = \pi$ and $\Delta U = M\,g\,l$. If the moment of inertia about an axis parallel to the rotation axis and passing through O is known, then the moment of inertia about P can be obtained by making use of the parallel axis theorem. If the angular velocity is reduced to zero, the static equilibrium condition corresponds to a minimum in the potential energy for $\theta = 0$ and the centre of mass is at its lowest point.

6.6 ROLLING MOTION

The rolling motion of cylindrical objects, such as wheels on vehicles, provides an interesting and important example of rotational motion in the Earth's gravitational field. A rolling object is in contact with a supporting surface, which provides an upward reaction force equal in magnitude to the downward weight force. If the rotating object has constant velocity, with no slippage, then it follows that there is a *static* friction force acting at the point of contact of the cylinder with the supporting surface. This is because at all times the small contact region of the rotating cylinder has zero velocity

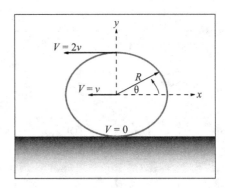

FIGURE 6.8   End view of a cylinder of radius R undergoing rolling motion, without slipping, on a horizontal surface. The centre of mass of the cylinder moves in the $-x$ direction with speed v. Points on the surface of the cylinder have instantaneous horizontal velocities V which depend on their orientation θ. The speed at the cylinder's point of contact with the surface is zero.

with respect to the surface. Static friction prevents slippage and produces a torque on the cylindrical object, causing it to roll. Friction plays an important role when the angular velocity of the cylinder is either increasing or decreasing.

Figure 6.8 shows the velocity of various parts of the outer surface of a rolling cylinder of radius R at some instant. Viewed in a frame of reference attached to the Earth's surface, the whole cylinder moves with constant velocity v in the $-x$ direction with the speed given by $v = 2\pi R/T = \omega R$ where T is the time taken for one revolution of the cylinder. Note that the cylinder rotates in a counterclockwise sense.

If the centre point of the cylinder moves horizontally with constant speed v, as shown, then the instantaneous horizontal speed for a point on the cylinder's surface is given by $V = v + \omega R \sin \theta = v(1 + \sin \theta)$ where θ is measured from the $+x$ direction as shown in Figure 6.8. For $\theta = \pi/2$, at the cylinder's top point, the horizontal speed is $V = 2v$, while at the bottom the speed is $V = 0$ for $\theta = 3\pi/2$. Since $V = 0$ at the cylinders contact point with the surface, there is a static friction force acting at this interface as stated above. In accelerated motion of the cylinder, friction plays an important role as the angular velocity of the wheel changes. If slippage does not occur, the friction force depends on μ_s, the static friction coefficient.

Exercise 6.7: A high-performance Formula One race car reaches a speed of 220 km/h along a straight length of track. If the wheels have a diameter of 46 cm, determine their angular velocities. Determine the instantaneous speed of points at the top and bottom of a wheel. Compare the speed obtained for the top of the wheel with the speed of sound in air.

The angular velocity is obtained using the circular motion relationship,
$$\omega = v/R = \frac{(220 \times 1000/3600)}{0.23} = 266\,\text{rad/s}.$$ The speed at the top of the wheel

is $v_{top} = v + \omega R = 2v = 532$ m/s. The speed at the bottom point, where the tyre meets the track, is $v_{bottom} = v - \omega R = 0$ m/s. The speed of sound in air is approximately 346 m/s. It follows that the speed of the top of the wheel exceeds the speed of sound in air.

Exercise 6.8: A solid cylinder of mass M and radius R rolls with a constant centre of mass velocity v on a flat horizontal surface as illustrated in Figure 6.9. Obtain an expression for the kinetic energy of the cylinder in terms of M and v. Next, if the surface of length L is tilted through an angle θ, what is the speed of the cylinder at the bottom of the incline after rolling without sliding down the slope starting from rest at the top?

The kinetic energy of the rolling cylinder involves both translational and rotational contributions and has the form $K = \frac{1}{2}M v^2 + \frac{1}{2}I \omega^2$, with I the moment of inertia about the axis of rotation and ω the angular velocity. Substituting $I = \frac{1}{2}M R^2$ and using $\omega = v/R$, gives $K = \frac{1}{2}M v^2 + \frac{1}{2}\left(\frac{1}{2}M R^2\right)\left(\frac{v}{R}\right)^2 = \frac{3}{4}M v^2$.

The rotational kinetic energy, $\frac{1}{4}M v^2$, is equal to one third of the total kinetic energy of a rolling solid cylinder.

For the inclined surface case, it is convenient to make use of mechanical energy conservation $\Delta K + \Delta U = 0$, where ΔK and ΔU are the changes in kinetic

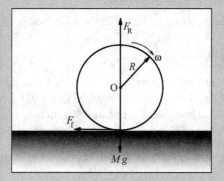

FIGURE 6.9 A cylinder of mass M and radius R rolls along a horizontal surface with angular velocity ω. The forces acting on the cylinder are the weight $M g$, the reaction F_R, and the friction F_f. For constant-speed rolling motion, the friction force at the point of contact with the surface is zero.

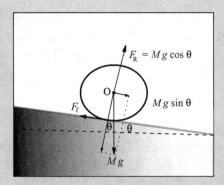

FIGURE 6.10 End view of a cylinder of radius R and mass M which rolls without slipping down an inclined plane. The weight $M\,g$ is resolved into components $M\,g\sin\theta$ parallel to the plane and $M\,g\cos\theta$ perpendicular to the plane. The reaction force F_R and the static friction force F_f act as shown. The torque produced by F_f about the rotation axis through O causes the cylinder to roll.

energy and potential energy, respectively. The Earth's gravitational field exerts a downward force on the cylinder, and the component of this force down the incline produces acceleration. Mechanical energy is conserved because the friction force is static at the point of contact of the cylinder with the ramp. No sliding motion of the cylinder occurs. The forces acting on the cylinder are shown in Figure 6.10. Note that the friction force acts *up* the incline and the torque it exerts about the axis of rotation produces the clockwise rolling motion.

The change in potential energy is $\Delta U = -M\,g\,L\sin\theta = -M\,g\,h$ where h is the vertical height through which the centre of mass of the cylinder effectively falls. Allowing for both translational and rotational energy using the expression for K obtained above gives the change in total kinetic energy as $\Delta K = \frac{3}{4}M\,v^2$.

It follows that $\frac{3}{4}M\,v^2 = M\,g\,h$ and $v = \sqrt{\frac{4}{3}g\,h}$.

Further insight into rolling motion is obtained by using Newton's second law for translational and rotational motions. In terms of the symbols shown in Figure 6.10, the second law gives for translational motion $M\,g\sin\theta - F_f = M\,a$ and for rotational motion $F_f\,R = I\,\alpha$. Combining these two equations to eliminate F_f, and using $I = \frac{1}{2}M\,R^2$ for a uniform cylinder, together with $a = \alpha R$, gives $g\sin\theta = \left(\frac{1}{2}+1\right)a = \frac{3}{2}a$. The translational acceleration down the inclined plane is constant, and the final velocity squared at the bottom can be obtained using the third kinematic equation. This procedure gives $v^2 = 2a\,L = 2\times\left(\frac{2}{3}g\sin\theta\right)\times L = \frac{4}{3}g\,h$, which agrees with the expression obtained above using mechanical energy conservation. Note that, because of differences in the moment of inertia expressions of objects, the value of v obtained for a uniform cylinder has a different numerical factor to those which would apply to a rolling hoop or a solid sphere.

6.7 STATIC EQUILIBRIUM

In Section 6.2, it is stated that the necessary conditions for equilibrium of a mechanical system are that both the forces $\mathbf{F}_i$ and the torques $\mathbf{\Gamma}_i$ acting on the object of interest must sum to zero, i.e. $\sum_i \mathbf{F}_i = 0$ and $\sum_i \mathbf{\Gamma}_i = 0$. These conditions need to be extended to include the potential energy of the system in order to ensure stable equilibrium. Consider, for example, a situation in which a body has an axis of rotation that does not pass through the centre of gravity. It is possible, although difficult in practice, to achieve a situation of unstable equilibrium in which the centre of mass lies vertically *above* the axis of rotation. A slight movement of the object will result in rotation towards a new stable equilibrium situation, with the centre of gravity directly below the axis of rotation. The gravitational potential energy of the system decreases in this process. Stable equilibrium therefore requires that the system be in a configuration with minimum potential energy. Following a small displacement away from stable equilibrium, the system will return to the stable configuration much as a round object will settle in the bottom of a hemispherical bowl. Note that if the axis of rotation for an object passes through its centre of gravity, then all orientations of the object are stable. In many cases, the minimum potential energy condition can readily be seen to be satisfied without the need for calculation.

In applying the static equilibrium conditions to particular situations, it is convenient to introduce two terms that are commonly used in dealing with static torques. These terms are firstly the *lever arm* and secondly the *moment* of a force about some chosen point. The significance of these terms is explained with the aid of Figure 6.11.

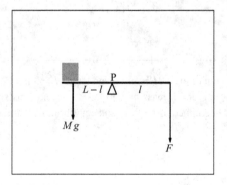

FIGURE 6.11 A light rod of length L is supported at P as shown. The gravitational force on mass M produces a counterclockwise moment of magnitude $M\,g\,(L-l)$ about the pivot point P. In equilibrium, the clockwise moment $F\,l$ produced by the force F with lever arm l balances the system. The reaction force at the support at P produces no moment around this point.

For a system in equilibrium, the sum of the moments around any point is zero. In applying this moment rule to a situation involving several forces, it may be convenient to calculate moments around a point through which an unknown force, such as a frictional force, acts in order to facilitate calculations as illustrated in Exercise 6.9.

Exercise 6.9: A solid steel beam of mass M = 50 kg and length 3 m rests on supports located 0.5 m from each end. Upward reaction forces F_1 and F_2 are exerted on the beam by the supports designated 1 and 2. An object of mass m = 40 kg is suspended 1.0 m from end 2. Determine the forces F_1 and F_2 using the static equilibrium conditions.

The supported beam arrangement is sketched in Figure 6.12. Taking moments around point 2, with clockwise moments as positive, leads to the relationship $F_1 \times 2 - M g \times 1.0 - m g \times 0.5 = 0$. Substituting the values for M and m with $g = 9.8 \, \text{m/s}^2$ gives $2F_1 = 9.8 \times (50 \times 1.0 + 40 \times 0.5) = 686 \, \text{kg m}^2/\text{s}^2$, and hence $F_1 = 343 \, \text{N}$.

Next, balancing the vertical forces gives $F_1 + F_2 - (M + m) g = 0$, and using the value obtained for F_1 gives $F_2 = 539 \, \text{N}$. Note that $F_2 > F_1$ because mass m is closer to end 2 than to end 1.

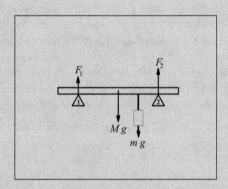

FIGURE 6.12   A beam of mass M = 50 kg and length L = 3 m is supported by trestles at 0.5 m from each end, and a mass m = 40 kg is suspended as shown. The reaction forces F_1 and F_2 are determined using the static equilibrium conditions.

Exercise 6.10: If the 50 kg steel beam, together with its suspended object of mass 40 kg, as described in Exercise 6.9, is removed from its supports, and is allowed to lean against a smooth vertical wall, determine the forces on the beam produced by the wall at the end labelled 1, and by the horizontal surface on which end 2 rests. The beam makes an angle $\theta = 63°$ with the horizontal surface. Assume that the *frictional* force at the contact point with the wall is negligibly small.

Figure 6.13 shows the forces acting on the beam of length $L = 3$ m. In addition to the two downward weight forces, shown as $M\,g$ at the centre of gravity, located at a distance $L/2$ from end 2, and $m\,g$ at a distance $l = 1$ m from end 2, there are three unknown forces which are the reaction forces, F_{1R} normal to the wall and F_{2R} normal to the horizontal surface, plus the horizontal friction force F_{2f}.

With three unknown forces to be determined, it is necessary to write down three equations, based on the static equilibrium conditions, which relate the forces. Firstly, balancing horizontal forces gives $F_{1R} - F_{2f} = 0$, while secondly, balancing vertical forces gives $F_{2R} - M\,g - m\,g = 0$, and thirdly, taking moments about end 2 leads to $F_{1R}\,L\sin\theta - M\,g\left(\dfrac{L}{2}\right)\cos\theta - m\,g\,l\cos\theta = 0$. Inserting numerical values gives $F_{1R} = 191$ N using the third equation and $F_{2R} = 882$ N from the second equation. The first equation, with the value obtained for F_{1R}, leads to $F_{2f} = 191\,$N.

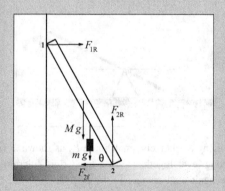

FIGURE 6.13 In addition to its known weight $M\,g$ and that of the suspended object $m\,g$, the canted beam experiences the reaction forces F_{1R} and F_{2R} together with a friction force F_{2f}. The reaction and friction forces are determined using the static equilibrium conditions.

6.8 THE LEVER

The use of the mechanical lever arm advantage provides a practical application of the zero-torque condition for equilibrium of an extended object. By making use of a long lever arm, a relatively small force F_1 applied at one end of the lever can balance a much larger force F_2 at the other end, provided a stable support, called a fulcrum, is positioned close to the end at which the large force acts as illustrated in Figure 6.14.

Consider the system shown in Figure 6.14 to be in a static equilibrium state with the mass M suspended just above the surface on which it previously rested. To simplify matters, only the vertical components of the forces acting on the two ends of the lever are shown. In addition, the large vertical force exerted by the fulcrum on the lever is omitted. In equilibrium, the moments of the forces about the fulcrum are equal in magnitude but opposite in sign giving $F_1 L_1 = F_2 L_2$ where L_1 and L_2 are the lever arms associated with the forces. The ratio L_1/L_2 is referred to as the mechanical advantage, which can be made much larger than unity by using a long lever on a robust fulcrum.

Levers can be used to move large objects with the application of relatively small forces. When raising the mass M through a small upward vertical distance h in the Earth's gravitational field, the potential energy increases by $\Delta U = M g h$. Mechanical energy conservation requires that $\Delta U = \Delta W = F_1 l$, with l the downward vertical distance moved by the other end of the lever. Using the properties of similar triangles gives that the mechanical advantage is the ratio $l/h = L_1/L_2$, as indicated above.

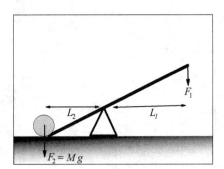

FIGURE 6.14 By applying a relatively small force F_1, as shown, a large object of mass M can be raised using a long lever together with a fixed rigid fulcrum. The greater the ratio L_1/L_2, the greater the mechanical advantage.

7 Fluids and Solids

7.1 INTRODUCTION

The properties of fluids and solids are important in a wide variety of technological applications. Solid materials are used in large mechanical structures, including bridges, buildings, and motor vehicles, as well as in small electronic devices such as computer chips. A deep understanding of a particular material requires microscopic measurements using special equipment, and complementary theoretical calculations. For many applications, much can be learnt about the mechanical properties of both solids and fluids using macroscopic measurements together with analysis based on classical mechanics. This approach is adopted in the present discussion.

As a starting point in this chapter, consider the phase diagram for a representative homogeneous substance plotted in terms of temperature and pressure as shown in Figure 7.1. The concepts of temperature and pressure are familiar from everyday experience. These quantities play key roles in determining the mechanical behaviour of materials. Pressure P is defined as the force per unit area acting on a surface, which may be the surface of a solid or the boundary of a vessel containing a fluid. In the SI system, the unit of pressure is defined as 1 N/m^2, which is called the pascal abbreviated as Pa. In plotting phase diagrams, it is advantageous to use the absolute temperature T scale, also known as the Kelvin scale, with kelvin units denoted by K. A related scale, called the Celsius scale, is used for many purposes in normal life. While the degree sizes are the same, with 1 K equivalent to $1\,°$C on the Celsius scale, the zero on the Kelvin scale, known as absolute zero or 0 K, occurs at $-273.16\,°$C. Temperatures in kelvins are of fundamental importance in examining the thermal properties of matter, as described in Chapters 11 and 12.

In Figure 7.1, the lines shown represent phase boundaries between the three phases, solid, liquid, and vapour, in which the system can be found. The physical characteristics in the three phases are quite different. In determining the mechanical properties, such as Young's modulus or the compressibility of a solid substance, as discussed below, it is usually necessary to work at constant temperature and pressure conditions, which may, of course, be ambient conditions, since the properties are generally temperature- and pressure-dependent. Details of phase diagrams for different materials can vary widely depending on the strength and nature of bonding at the atomic level. In particular, solids are broadly classified into four categories named

DOI: 10.1201/9781003485537-7

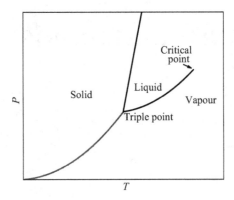

FIGURE 7.1   Representative phase diagram for a substance showing the solid, liquid, and vapour regions as a function of temperature T and pressure P. Transitions between the phases occur at the phase boundaries represented by lines in the diagram. At the triple point, all three phases coexist, while at the critical point and above the liquid and vapour phases become indistinguishable.

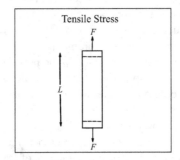

FIGURE 7.2   Side view of a solid rod subject to uniaxial stress produced by equal and opposite applied forces F. A small increase in the length occurs while the stress is applied.

molecular (e.g. solid argon and polymers), ionic (e.g. NaCl), covalent (e.g. silicon), and metallic (e.g. copper and silver) solids. Differences in the electronic structure are responsible for the differences in the binding energies of the atomic constituents, and these differences are reflected in the mechanical and other properties. There are other types of solid, such as alloys, glassy materials, and liquid crystals, which are not covered by the above classification.

The bulk mechanical properties of solid materials are characterized by their response, called the strain, to various types of applied stress. A uniaxial case is illustrated in Figure 7.2, which shows a 2D side view of a solid object under longitudinal stress. Equal and opposite forces F are applied to the top and bottom surfaces of the object, producing a small change in its length. Note that the forces are arranged to act uniformly over the surfaces to which they are applied. In practice, a long rod or wire specimen is typically used in measuring the strain produced by an applied stress.

Two other types of stress, called shear stress and hydrostatic pressure, produce shear strains and compressive strains, respectively. Further details and definitions of the corresponding elastic moduli are given in Section 7.5. Over a limited range, it is found that strain is proportional to applied stress for many solids. Furthermore, the strain is elastic in nature, with no permanent deformation after the stress is removed. In this *linear response* range, measurements of a particular stress–strain response for a specimen yield the corresponding elastic modulus for the material as explained in Section 7.5. Uniaxial and shear stress measurements cannot be made on fluids because the fluids respond by flowing until equilibrium is attained in the vessels in which they are contained. It is, however, possible to measure what is termed the compressibility by applying pressure to a fluid as shown in Section 7.2. The topic of fluid flow is dealt with in Section 7.4.

7.2 PRESSURE IN FLUIDS

7.2.1 PRESSURE AND DENSITY

When considering pressure effects in fluids, it becomes clear that the fluid density ρ is of particular importance. If a mass m of fluid occupies a volume V, then the density is $\rho = m/V$ with SI units of kg/m^3. The density depends on temperature and pressure, and it is therefore necessary to specify these conditions. For fluids, it is often convenient to quote ρ values measured at $0°C$ and 1 atmosphere (atm; 1 atm equals 101.325 kPa in SI units). These conditions are known as standard temperature and pressure (STP). For liquids, the variation in density is generally small for limited changes in temperature and pressure. This is not the case for gases, for which ρ has a strong dependence on T and P. Table 7.1 lists the values of ρ for several fluids.

For a liquid such as water, ρ is, to a good approximation, independent of depth for modest depths of the order of meters, and the liquid behaves as though it were incompressible. In contrast, the density of a gas such as air is strongly dependent on altitude measured from the Earth's surface. The drop in density with altitude is primarily responsible for the oxygen deficiency effects experienced by mountain climbers.

TABLE 7.1
Densities of Representative Fluids at STP

Substance	Density (kg/m³)
Water	1.000
Ethanol	0.80
Mercury	13.6
Air	1.29×10^{-3}
Helium	0.179×10^{-3}

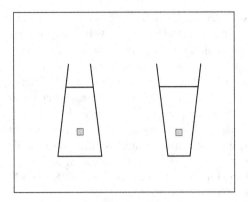

FIGURE 7.3   Volume elements (shown in grey), which are in static equilibrium in containers of different shapes, experience identical forces on their surfaces if they are at the same depth in the liquid.

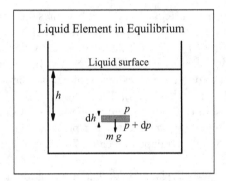

FIGURE 7.4   A representative 3D volume element, shown shaded in a side-on view, is of thickness dh with top and bottom surfaces of area A. The element is in equilibrium under the action of its downward weight and the net upward force due to the pressure difference between its upper and lower surfaces.

The pressure at a particular depth in a fluid is independent of the orientation of the surface on which the pressure acts. This can be understood by considering a volume element of fluid in equilibrium. The forces on the surfaces of the element must sum to zero or the element would move until the forces did balance. In addition, the pressure is independent of the shape of the container as illustrated in Figure 7.3.

7.2.2   VARIATION OF PRESSURE WITH DEPTH IN AN INCOMPRESSIBLE LIQUID

Consider a cylindrical volume element of cross-sectional area A and thickness dh in a column of incompressible liquid as depicted in Figure 7.4. The liquid is situated in the Earth's gravitational field, which produces a downward weight force on the element that leads to a pressure increase with depth in the fluid as shown below.

It is convenient to choose the h-axis to point vertically downwards with the origin located at the liquid's surface. Atmospheric pressure at the liquid's surface is p_0. If the pressure at the upper surface of the volume element is p, then the pressure at the lower surface is $p + dp$, with the difference in pressure due to the need to support the weight of liquid $w = \rho A g \, dh$ in the element.

Balancing forces on the volume element leads to the equation $(p + dp)A = p A + \rho A g \, dh$, which gives $dp = \rho g \, dh$. In order to determine the pressure at depth h in the liquid, this equation is integrated as follows: $\displaystyle\int_{p_0}^{p} dp = \int_{0}^{h} \rho g \, dh$

or $p - p_0 = \rho g h$. After rearrangement, the following result is obtained:

$$p = p_0 + \rho g h \qquad (7.1)$$

Equation (7.1) shows that the pressure in a liquid increases linearly with depth h below the liquid's surface. The pressure p_0 at the surface is typically atmospheric pressure.

Exercise 7.1: What is the pressure at the bottom of a freshwater lake of depth 8 m? Take atmospheric pressure as 1 atm = 1.01×10^5 Pa and the density of the lake water as 10^3 kg/m^3.

Using Equation (7.1), the pressure at the bottom of the lake is given by $p = 1.01 \times 10^5 + 10^3 \times 9.8 \times 8 = (1.01 + 0.78) \times 10^5 = 1.79 \times 10^5$ Pa. The pressure at the lake bottom is slightly less than 2 atm, with almost half contributed by the lake water and the remainder by the atmosphere above the lake.

7.2.3 VARIATION OF PRESSURE WITH ALTITUDE IN A COMPRESSIBLE GAS

Consider a finite column of air stretching from the Earth's surface to an altitude of a few hundred meters, which is the height of fairly low cloud cover. The altitude limitation is introduced to allow the following simplifying assumptions to be made: firstly, that the temperature of the gas is roughly the same throughout the column, and secondly, that any variation in g can be neglected. For a compressible gas, the density is not constant but varies with position in the column. To a good approximation, a volume element containing a constant number of molecules in the air column obeys the ideal gas equation of state $PV = nRT$ with P the pressure, T the absolute temperature, n the number of gas molecules in volume V, and R the gas constant. Since T is assumed to be constant, it follows that $P \propto n/V$, and therefore $P \propto \rho$ with $\rho = n/V$ being the *molar* density. Let the pressure at the bottom of the column be P_0 while the density at this level is ρ_0. It follows that $\rho/\rho_0 = P/P_0$ where ρ and P are, respectively, the density and pressure at height h.

Now consider a thin horizontally oriented slice of air of thickness dh at height h in the column. Taking the air density to be constant in the small slice leads to the following expression for the pressure difference between the top and bottom surfaces of the slice $dP = -\rho g\, dh = -\left(\rho_0 \dfrac{P}{P_0}\right) g\, dh$ where use has been made of the result

$\rho/\rho_0 = P/P_0$ for the pressure and density ratio as given above. The minus sign indicates

that P decreases as h increases. Rearranging and forming integrals gives $\displaystyle\int_{P_0}^{P} \dfrac{dP}{P} = -\dfrac{\rho_0 g}{P_0}\int_{0}^{h} dh.$

Integration leads to the relationship $\ln\left(\dfrac{P}{P_0}\right) = -\dfrac{\rho_0 g h}{P_0}$. Taking antilogarithms yields

the required result:

$$P = P_0 \exp\left(-\dfrac{\rho_0 g h}{P_0}\right) \tag{7.2}$$

Equation (7.2) shows that in the air column the pressure falls off exponentially with altitude for fairly low altitudes. The assumptions made start to break down as the altitude increases.

Exercise 7.2: The air pressure on a mountain top is 0.9 atm. Estimate the height of the mountain top above sea level. Take the density of air at sea level and at ambient temperature as 1.21 kg/m^3. 1 atm = 1.01×10^5 Pa.

Assuming that Equation (7.2) holds over the altitude range, and taking logarithms of both sides gives $h = \dfrac{P_0}{\rho_0 g}\ln\left(\dfrac{P_0}{P}\right) = \dfrac{1.01\times 10^5}{1.21\times 9.8}\ln\left(\dfrac{1}{0.9}\right) = 897\,\text{m}.$

7.2.4 COMPRESSIBILITY OF FLUIDS

The isothermal compressibility of a material, denoted by κ, relates the fractional change in volume of the material to an increase in pressure as expressed in the relationship $\Delta V/V = -\kappa\,\Delta P$. The volume decreases with the increase in pressure, and the minus sign is therefore inserted to give positive values for κ. In terms of infinitesimal changes, the isothermal compressibility is defined as

$$\kappa = -\dfrac{1}{V}\dfrac{dV}{dP} \tag{7.3}$$

The units for κ are Pa^{-1}. Typical gas compressibility values are much larger than those of liquids. That is because the molecules in a gas are on average well separated, while in liquids they are packed more closely together. For air and many other gases, which closely obey the ideal gas equation of state under normal conditions of temperature and pressure, the compressibility value is $\kappa \approx 10^{-5}\,\text{Pa}^{-1}$. Water at room temperature has a κ value roughly four orders of magnitude smaller than that of air, while solids have even smaller κ values. In dealing with the mechanical properties of solids, it is usual to introduce the bulk modulus, which is the reciprocal of the compressibility, as discussed in Section 7.5.

Exercise 7.3: Obtain an expression for the isothermal compressibility of a gas that obeys the ideal gas equation of state $PV = nRT$.

From the ideal gas equation $\dfrac{dV}{dP} = -\dfrac{nRT}{P^2} = -\dfrac{V}{P}$, and using Equation (7.3) this gives $\kappa = 1/P$.

Taking atmospheric pressure as 1.01×10^5 Pa at sea level, it follows that $\kappa \sim 10^{-5}$ Pa^{-1} as given above. The compressibility of air increases with altitude due to the decrease in its pressure.

7.3 FLUID STATICS

7.3.1 THE PRINCIPLES OF HYDROSTATICS

Using the expressions for the pressure in a fluid given in Section 7.2, and, in particular the variation of pressure with depth in a liquid, it is possible to explain a variety of phenomena and to develop simple but useful applications, including buoyancy devices, mercury barometers, liquid siphons, and the hydraulic press. Historically, two famous principles were put forward based on observations made of the behaviour of liquids. They are known as Archimedes' principle and Pascal's principle, respectively.

Archimedes' principle states that *a body wholly or partly immersed in a fluid experiences a buoyancy force equal to the weight of fluid displaced.*

Pascal's principle states that *the pressure exerted on an incompressible fluid is transmitted evenly throughout the fluid.*

These principles were implicitly assumed in deriving the relationship $p = p_0 + \rho g h$, which is given in Equation (7.1), for the pressure variation as a function of depth in an incompressible liquid. At depth h below the surface, the pressure depends on, firstly, p_0, the atmospheric pressure at the surface, and secondly, the weight of a fluid column of height h and unit cross-sectional area A. An increase in p_0 leads to an increase in p at every point in the liquid in accordance with Pascal's principle. If the liquid column of height h were replaced by a solid object with precisely the same dimensions as those of the column, the object would experience the same upward force as that experienced by the liquid column. This buoyancy force acts in accordance with Archimedes' principle. It follows that an object of any shape would experience an upward force equal to the weight of fluid displaced, with the force acting through the centre of gravity of the object. If the density of the object is greater than the density of the fluid, the object will sink to the bottom of the container. Conversely, if the density of the object is less than that of the liquid, the object will rise to the surface and float.

Exercise 7.4: Determine the fraction of the volume of an iceberg that is submerged below the surface of the sea in which it floats. Take the density of ice as $\rho_I = 0.92 \times 10^3$ kg/m³ and that of seawater as $\rho_{sw} = 1.025 \times 10^3$ kg/m³.

From Archimedes' principle, the buoyancy force on the floating iceberg of volume V is given by $\rho_{sw} \, V_{sw} \, g$ where V_{sw} is the volume of seawater *displaced* by the submerged portion of the iceberg. In equilibrium, the buoyancy force is equal to the iceberg's weight so that $\rho_{sw} \, V_{sw} \, g = \rho_I \, V \, g$. The submerged volume fraction of the iceberg is therefore $V_{sw}/V = \rho_I/\rho_{sw} = 0.92/1.025 = 0.90$. Ninety percent of the iceberg's volume is submerged in the sea.

7.3.2 Pressure Measurement

Pressure-measuring devices based on the relationship $p = p_0 + \rho \, g \, h$ given in Equation (7.1) include manometers and barometers. A simple manometer is as illustrated in Figure 7.5.

Manometers are typically made of a transparent material, such as glass, shaped in the form of a U-tube, with one end connected to a vessel in which the gas pressure is to be measured and the other end open to the atmosphere. The U-tube contains an inert liquid, such as mineral oil or mercury, of known density ρ_0. When the pressure in the vessel is equal to atmospheric pressure the two surface levels of the manometer liquid are at the same height above a chosen reference level. As the pressure in the vessel is varied, the liquid surface levels move up or down. The vertical height difference h between the levels in the two arms of the manometer gives a measure of the pressure difference $\Delta p = p - p_0$ between the pressure of the gas in the vessel, designated p, and that of the local atmosphere, designated p_0, via the relationship $\Delta p = \rho \, g \, h$. Note that h may be positive or negative depending on whether the gas pressure in the vessel is larger or smaller than atmospheric pressure. Pressures measured using manometers, or other devices that measure pressure differences with respect to atmospheric pressure, are called *gauge* pressures. Gauge pressure can be

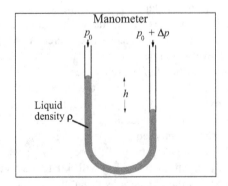

FIGURE 7.5   Manometer for measuring gauge pressure Δp, which may be above or below atmospheric pressure p_0. Hydrostatics gives $\Delta p = \rho g h$.

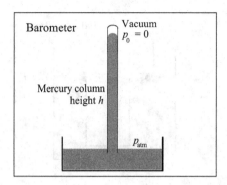

FIGURE 7.6 Mercury barometer for measuring atmospheric pressure p_{atm}. In a practical instrument, the mercury bath would be enclosed to prevent mercury vapour from polluting the atmosphere.

converted to the actual pressure in Pa by adding or subtracting atmospheric pressure when required.

The pressure of the atmosphere near the Earth's surface is measured using instruments called barometers. The mercury barometer is a classic example of this type of instrument and is again based on the relationship between pressures at different heights in a fluid as given in Equation (7.1). Mercury barometers make use of the high density of mercury, $\rho_{Hg} = 13.6 \times 10^3$ kg/m³, to keep the device fairly compact. A column of the liquid is contained in a long tube, the lower end of which dips into a container of mercury that is open to the atmosphere. The upper end of the tube is sealed, with the space above the liquid evacuated. A simple barometer is shown in Figure 7.6.

From Equation (7.1), the pressure of the mercury in the vertical tube is given by $p = p_0 + \rho_{Hg} g h$ with h the vertical height of the mercury meniscus above the level of the surface in the container. Since the space above the mercury in the tube is evacuated, the pressure $p_0 = 0$. Also, $p = p_{atm}$ because fluid pressures at the same level are equal. Atmospheric pressure is therefore given by $p_{atm} = \rho_{Hg} g h$. A pressure of one atmosphere supports a mercury column of height 0.76 m.

In addition to the mercury barometer, a number of mechanical barometers for measuring air pressure without the use of fluids have been developed. A well-known device is the aneroid barometer, which involves an evacuated cell made of thin flexible metal. Changes in pressure are detected by the force exerted on a spring attached to the cell. Very small microelectromechanical system (MEMS) devices have been incorporated into cell phones.

7.3.3 THE HYDRAULIC PRESS

The hydraulic press is an important practical device based on Pascal's principle. Consider two cylinders, with cross-sectional areas A_1 and A_2 respectively, connected together by a pipe. The system is partially filled with a hydraulic liquid which is

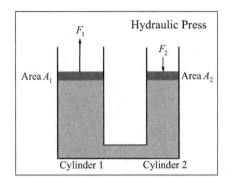

FIGURE 7.7   Hydraulic press with two connected piston-cylinder chambers. The chambers are filled with a hydraulic liquid which is effectively incompressible. Force F_1 is larger than F_2 by the ratio of the piston areas A_1/A_2.

effectively incompressible. The cylinders are fitted with pistons to which weights can be added to achieve the desired initial static equilibrium state. The application of an additional downward force F_2 to piston 2 produces an increase in the upward force F_1 on piston 1 as shown in Figure 7.7.

The downward force F_2 on piston 2 causes it to move downwards through a distance d_2. In order to keep the total volume of the incompressible liquid constant, the piston in cylinder 1 moves *upwards* through a distance $d_1 = \left(\dfrac{A_2}{A_1} \right) d_2$. Using Pascal's principle, the pressure increase Δp in the system produced by F_2 is transmitted throughout the fluid, and this leads to the relationship $\Delta p = F_1/A_1 = F_2/A_2$, and hence $F_1 = \left(\dfrac{A_1}{A_2} \right) F_2$. For $A_1 > A_2$, it follows that $F_1 > F_2$. Thus, a small downward force on piston 2 is converted to a much larger upward force on piston 1 through the action of the press, which serves as a hydraulic lever. It is illuminating to consider the work done by the forces. For F_1, the work done is $W_1 = F_1 \, d_1$, and that done by F_2 is $W_2 = F_2 \, d_2$. Substituting for F_1 and d_1 using the relationships $F_1 = \left(\dfrac{A_1}{A_2} \right) F_2$ and $d_1 = \left(\dfrac{A_2}{A_1} \right) d_2$ given above, shows that $W_1 = W_2$. The large upward force F_1 acts through a much smaller distance than the downward F_2 does, so that the work done by the two forces is the same, as required by mechanical energy conservation assuming friction forces are negligible.

7.4   FLUID FLOW

In dealing with the flow of fluids, it is important to distinguish between turbulent flow and non-turbulent flow. Turbulent flow involves time-varying flow patterns, while in non-turbulent flow the patterns are time-independent. Because of the complexity of analysing turbulent flow, the present discussion will be limited to non-turbulent flow. Furthermore, only incompressible fluids will be considered. In discussing flow patterns, it is useful to consider the paths followed by small fluid elements. For steady non-turbulent flow, the paths are represented by what are called streamlines and the flow is called streamline flow. Experimental techniques involving the introduction of traces of dye into a flowing fluid have been developed in order to observe and photograph streamline patterns. As an illustration, consider the flow of a liquid through a tube in which there is a constriction as shown in Figure 7.8. Note that in the region of the constriction the streamlines are brought closer together, and this corresponds to an increase in the fluid flow velocity as discussed below. It is additionally necessary to distinguish between viscous flow and non-viscous flow. This distinction involves the magnitude of the viscosity coefficient η, which is introduced in Section 7.4.3. It is convenient to start by considering non-viscous flow, especially as water has a low viscosity coefficient and therefore satisfies the requirements for this type of flow. Viscous flow is discussed separately.

7.4.1 THE CONTINUITY EQUATION

For an incompressible fluid undergoing a flow process, the density ρ is constant at all points in the fluid. In contrast, the velocity of fluid elements and the local pressure may vary with position, as shown below in Section 7.4.2, which introduces Bernoulli's equation. The continuity equation, given in Equation (7.4) below, expresses the requirement that the mass of fluid per second entering a tube of flow is equal to the mass per second leaving the tube. Consider fluid flow through a section of a tube and let the cross-sectional areas be A_1 at the entry to the section and A_2 at the exit as illustrated in Figure 7.8.

In time interval Δt, the volume of fluid entering the tube with speed v_1 is $\Delta V = v_1 A_1 \Delta t$, which has mass $\Delta m = \rho v_1 A_1 \Delta t$. The mass leaving the tube is given by $\Delta m = \rho v_2 A_2 \Delta t$. Equating the right-hand sides of these equations for the masses entering and leaving the tube gives the equation of continuity as

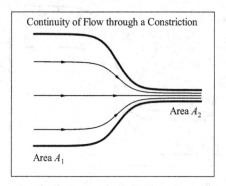

FIGURE 7.8 Non-turbulent fluid flow through a section of a pipe in which the cross-sectional area decreases from A_1 to A_2. The flow rates in the wide and narrow sections are related by the continuity equation.

$$A_1 v_1 = A_2 v_2 \tag{7.4}$$

The smaller the cross-sectional area becomes along the tube of flow, the higher the fluid speed in that region for a given flow rate, and, conversely, the larger the area the lower the speed. Note that the element of fluid with volume ΔV entering the tube of flow is effectively transmitted in time Δt from the entrance to the exit by the intermediate fluid in the tube. This follows because all fluid elements of a given geometry are identical to one another. In order to determine the variation of fluid pressure along a tube of flow it is necessary to use Bernoulli's equation, which is based on the work–energy theorem.

7.4.2 BERNOULLI'S EQUATION

Bernoulli's equation is derived for non-viscous, non-turbulent, incompressible fluid flow through a pipe. Referring to Figure 7.9, which shows a section of a pipe through which flow occurs, let the pressure acting on the area A_1 at the entrance to the pipe be p_1 and that on area A_2 at the exit be p_2. The corresponding forces are $p_1 A_1$ in the direction of flow and $p_2 A_2$ opposing the flow. The shaded regions in Figure 7.9 represent fluid elements entering and leaving the pipe during a time interval Δt.

In time Δt, the work done in transporting a small volume $\Delta V = A_1 v_1 \Delta t$ with mass $\Delta m = \rho \Delta V$ into the pipe section of interest is $W_1 = p_1 A_1 v_1 \Delta t$. In the same time interval, the work done by the force $p_2 A_2$ as the equivalent fluid element leaves the pipe is $W_2 = -p_2 A_2 v_2 \Delta t$. The net work done in this process is $W = p_1 A_1 v_1 \Delta t - p_2 A_2 v_2 \Delta t = (p_1 - p_2) \Delta V$. Because non-viscous flow is assumed, the work done is equal to the change in energy of the system associated with the effective transport of the mass Δm from the input to the output of the tube of flow. The

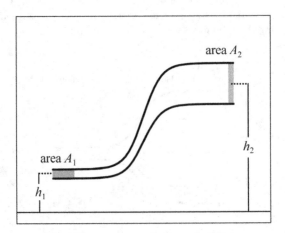

FIGURE 7.9   Fluid flow through a section of a pipe in which the cross-sectional area changes from A_1 to A_2. In addition, the fluid is transported upwards through height h in the Earth's gravitational field. The shaded segments represent equal volume elements of fluid entering and leaving the pipe section in a time interval Δt.

energy change is given by the sum of the change in kinetic energy $\Delta K = \frac{1}{2}\Delta m\left(v_2^2 - v_1^2\right)$ and the change in potential energy $\Delta U = \Delta m\,g\left(h_2 - h_1\right)$, where h_1 and h_2 are, respectively, the initial and final heights above a reference level of the centre of mass of Δm. Using $W = \Delta K + \Delta U$ gives $\left(p_1 - p_2\right)\Delta V = \frac{1}{2}\Delta m\left(v_2^2 - v_1^2\right) + \Delta m\,g\left(h_2 - h_1\right)$. Dividing through by ΔV and collecting terms leads to Bernoulli's equation in the form

$$p_1 + \frac{1}{2}\rho v_1^2 + \rho g h_1 = p_2 + \frac{1}{2}\rho v_2^2 + \rho g h_2 \qquad (7.5)$$

It is important to recognize that the quantities p, $\frac{1}{2}\rho v^2$, and $\rho g h$ which are involved in Bernoulli's equation have units J/m^3 and correspond to distinct contributions to the total energy density in a flowing fluid. It follows from Equation (7.5) that the total energy density remains constant for incompressible fluids in non-turbulent flow processes. An increase in one of the terms is therefore accompanied by a corresponding decrease in one, or both, of the other terms. Bernoulli's equation expresses the law of mechanical energy conservation in a convenient form. Together with the continuity equation, Bernoulli's equation provides a quantitative description of the flow behaviour of liquids, such as water, which approximate an ideal, non-viscous, incompressible fluid.

Exercise 7.5: A horizontal water pipe has an initial diameter of 4 cm that tapers gradually over a central section to a final diameter of 2 cm. If the pressure and speed of flow through the wide portion of the tube are, respectively, 2×10^4 Pa and 1 m/s, determine the pressure and flow speed in the narrow portion. The density of water is 10^3 kg/m^3.

Let the radii of the wide and narrow portions of the pipe be R_1 and R_2 respectively. The cross-sectional area of the wide portion of the pipe is then given by $A_1 = \pi R_1^2$ and that of the narrow portion by $A_2 = \pi R_2^2$. Taking the water velocity in the wide portion as v_1 and that in the narrow portion as v_2, the continuity equation gives $v_2 = \left(\dfrac{A_1}{A_2}\right)v_1 = \left(\dfrac{4}{1}\right)\times 1 = 4$ m/s.

Since the flow is horizontal, Bernoulli's equation simplifies through cancellation of the $\rho g h$ terms to give $p_1 + \frac{1}{2}\rho v_1^2 = p_2 + \frac{1}{2}\rho v_2^2$. Rearranging and substituting values gives $p_2 = p_1 + \frac{1}{2}\rho\left(v_1^2 - v_2^2\right) = 2\times 10^4 + \frac{1}{2}\times 10^3 \times\left(1-16\right) = 1.25 \times 10^4$ Pa.

Exercise 7.6: If the narrow portion of the pipe in Exercise 7.5 were bent upwards, and then bent back to the horizontal at a height of 50 cm above the original height, determine the water pressure in the raised horizontal section of the tube.

The flow velocity in the raised section of the narrow tube remains unchanged at 4 m/s as required by the continuity equation for an incompressible fluid, while the pressure in this section does change in accordance with Bernoulli's equation, which becomes $p_2' = p_1 + \frac{1}{2}\rho(v_1^2 - v_2^2) + \rho g(h_1 - h_2)$ where $h_2 - h_1 = 0.5\,\text{m}$. Note that p_1 and p_2' are now the pressures in the lower and upper sections of the narrow tube. Substituting numbers, and making use of the result of Exercise 7.5, gives $p_2 = 1.25 \times 10^4 - 10^3 \times 9.8 \times 0.5 = 7.6 \times 10^3$ Pa.

The pressure is lowered in the upper portion of the pipe. The kinetic energy of the flowing liquid does not change, since the flow velocity remains constant, but the potential energy does change with elevation in the gravitational field.

Many practical applications of Bernoulli's equation make use of the pressure drop in a constricted region of a tube through which fluid flows. Examples are the Venturi meter for measuring fluid flow rates through pipes, and spray atomizers which draw liquid up from a reservoir when a bulb is squeezed to force air through the device. Other examples of effects accounted for by Bernoulli's equation are the aerodynamic lift on the aerofoils of aircraft, and the swing in the trajectory of a spinning ball moving at speed through the air. In the latter examples, pressure differences arise on opposite sides of an object moving through a fluid when the fluid speed is higher on one side of the object than on the other.

7.4.3   VISCOUS FLUID FLOW

The discussion of fluid flow given above neglects viscous effects that dissipate energy in irreversible processes in a viscous fluid. While the continuity equation continues to hold, provided the fluid is incompressible, Bernoulli's equation breaks down for these fluids because mechanical energy conservation does not hold. In contrast to ideal fluids, for which no pressure drop is predicted in streamline flow through a horizontal pipe, viscous fluid flow is characterized by a pressure drop with distance along a horizontal pipe. The smaller the radius of the pipe, the greater the pressure drop over a given length. Investigations of the fluid velocity variation across a pipe through which viscous fluid flows have revealed that the flow velocity approaches zero at the walls of the pipe and is largest along the central axis of the pipe as illustrated in Figure 7.10.

Viscous fluid flow through a pipe can be viewed as a set of thin-walled concentric cylindrical tubes of flow moving parallel to the flow direction, with a radial decrease in velocity from the centre of the pipe to the wall as depicted in Figure 7.10, which shows the velocity profile in 2D across the pipe mid-section.

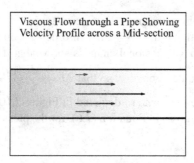

FIGURE 7.10 Viscous flow through a portion of a pipe, showing the velocity profile for a 2D cross-section of the pipe. In 3D, the flow pattern is visualized as concentric tubes of flow.

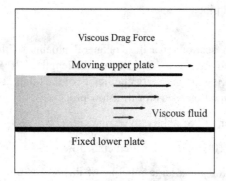

FIGURE 7.11 A viscous fluid (shown as the shaded region) is situated between a fixed lower plate and a moving upper plate, which experiences a drag force. There is a velocity gradient in the motion of successive fluid layers, as indicated by the arrow lengths.

In order to describe the viscous properties of fluids, it is necessary to introduce a quantity called the viscosity coefficient denoted by η. Consider an arrangement in which a fluid is contained between two flat horizontal plates of area A separated by a distance d. The upper plate is acted on by a horizontal force F which causes the plate to move at a constant velocity v with respect to the lower plate as illustrated in Figure 7.11.

Experiment shows that the force F necessary to produce steady motion of the upper plate with respect to the fixed lower plate is proportional to the product $A v$ and inversely proportional to d. The (dynamic) viscosity coefficient η is introduced as a proportionality constant to give $F = \eta \dfrac{A v}{d}$. The SI units of η are Pa s. Representative η values are approximately 1 mPa s for water at 20°C, and 100 mPa s for light oil at the same temperature.

For a viscous fluid flowing through a horizontal cylindrical pipe, there is a pressure drop in the fluid with distance along the pipe. Consider a pipe of length L and radius R through which a fluid with viscosity coefficient η flows at a volume flow rate Q.

The pressure drop Δp along the pipe is expected to depend on η, L, Q, and on the inverse of the cross-sectional area $A = \pi R^2$. In order to obtain the required units of Pa for the pressure drop, dimensional analysis shows that it is necessary to include a further factor R^2 in the denominator, so that $\Delta p = \dfrac{\Delta F}{A} \propto \dfrac{\eta L v}{A} \propto \dfrac{\eta L Q}{A^2}$. A detailed calculation based on the flow pattern depicted in Figure 7.10 leads to what is called Poiseuille's equation, as given in Equation (7.6), for the pressure drop in viscous fluid flow through a pipe:

$$\Delta P = \eta \frac{8L}{\pi R^4} Q \qquad (7.6)$$

Poiseuille's equation applies to the non-turbulent viscous flow of a fluid through a horizontal cylindrical pipe.

Exercise 7.7: A section of a horizontal oil pipeline with inside diameter 20 cm has a pressure drop of 120 kPa over its length of 10 km. Calculate the flow rate in L/s taking the viscosity coefficient of the oil as 200 mPa s. Assume that Poiseuille's equation holds for this flow process.

From Equation (7.6), the flow rate is $Q = \dfrac{1}{\eta} \dfrac{\pi R^4}{8} \dfrac{\Delta p}{L} = \dfrac{1}{0.2} \dfrac{\pi \times 10^{-4}}{8}$ $\dfrac{1.2 \times 10^5}{10^4} = 2.36 \times 10^{-3}$ m^3/s. The predicted flow rate is 2.36 L/s. The use of Poiseuille's equation for a large pipeline of this sort is questionable, and the calculated flow rate is significantly higher than would be achieved in practice.

7.5 MECHANICAL PROPERTIES OF SOLIDS

While liquids and gases occupy the space available to them, with allowance for the influence of gravitational forces, most solids have shapes and dimensions that change only slightly when subjected to moderate applied stress. Furthermore, stresses can be applied to solids in 1D and 2D, in addition to 3D. In dealing with fluids, a single property, the compressibility given by $\kappa = -\dfrac{1}{V} \dfrac{dV}{dP}$, is of key importance in determining the behaviour under stress. Gases are highly compressible, while liquids are not. Solids are, in general, even less compressible than liquids, and in the simplest case of a homogeneous isotropic material require two elastic constants to describe their mechanical properties. The elastic constants, or moduli, are called Young's modulus Y, and the shear modulus σ. Other elastic constants, including the bulk modulus B, can be defined, although their values depend on Young's modulus and the shear modulus. The bulk modulus and the compressibility are closely related as shown below.

The material dependent moduli for solids are defined using the general relation:

$$stress = modulus \times strain$$

where strain is a dimensionless relative deformation and stress has units of N/m^2. It follows that the moduli have the latter units as well.

The bulk modulus is determined by increasing the hydrostatic pressure on a solid specimen by Δp and measuring the resultant fractional volume change $\Delta V/V$. Thus, B is defined by the relationship

$$\Delta p = -B \frac{\Delta V}{V} \tag{7.7}$$

From Equation (7.7), it follows that $B = -V \frac{\Delta p}{\Delta V}$. Note that the bulk modulus B is the inverse of the compressibility κ.

Measurements of Young's modulus of a solid involve applying opposing forces F to each end of a specimen in the form of a long rod or wire of length L and cross-sectional area A. The tensile stress F/A produces a strain $\Delta L/L$ from which Y is obtained using the relationship

$$\frac{F}{A} = Y \frac{\Delta L}{L} \tag{7.8}$$

Shear modulus values are obtained by applying opposing forces F to the parallel upper and lower faces of a specimen of thickness L, as shown in Figure 7.12, to produce a shear stress, which results in a shear deformation ΔX as given in the following equation:

$$\frac{F}{A} = \sigma \frac{\Delta X}{L} \tag{7.9}$$

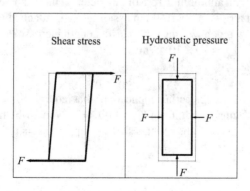

FIGURE 7.12 Representation in 2D of (a) shear stress and (b) hydrostatic pressure-induced strains for a solid specimen with a rectangular cross-section.

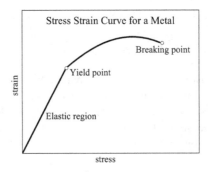

FIGURE 7.13   Stress–strain curve obtained for a representative stretched metal wire. Hooke's law is obeyed in the linear elastic region, where the slope of the line gives Young's modulus for the wire. For stress values above the yield point, the specimen undergoes permanent elongation. As the stress is increased, the permanent deformation increases until a breaking point is reached where catastrophic failure occurs.

Figure 7.12 gives a 2D representation of (a) shear stress and (b) hydrostatic pressure-induced strains in a rectangular solid. The case of uniaxial stress is illustrated in Figure 7.2.

The stress–strain curve for a metal wire typically shows a linear region as the stress is increased gradually from zero, with slope given by Young's modulus Y for the material as predicted by Equation (7.8). In the linear stress–strain region, the strain is reversible with the specimen returning to its original length when the applied stress is reduced back to zero. Stress–strain behaviour in the linear region is described by Hooke's law, which is discussed in Chapter 8. As the stress is increased further, the strain behaviour changes. It is found that above what is called the yield point the strain increases non-linearly and is no longer reversible. Permanent deformation of the specimen occurs. At sufficiently high stress values the specimen ruptures as indicated in Figure 7.13.

Exercise 7.8 An aluminium rod of diameter 2 mm and length 80 cm is suspended from a rigid support. If a mass of 50 kg is attached to the lower end of the rod, by how much will the rod's elastic length increase? Young's modulus for aluminium is $Y_{Al} = 70 \times 10^9$ N/m^2.

From Equation (7.8), $\Delta L = \dfrac{1}{Y}\dfrac{F}{A}L = \dfrac{1}{70 \times 10^9} \times \dfrac{50 \times 9.8}{\pi \times 10^{-6}} \times 0.8 = 1.8\,\text{mm}$. Steel, which is important in engineering applications that require tensile or compressive strength, has a Young's modulus of $Y_{Fe} = 200 \times 10^9$ N/m^2, which is significantly higher than that of some other metals such as copper ($Y_{Cu} = 110 \times 10^9$ N/m^2) or aluminium (Y_{Al}, given above).

As an example of shear strain, consider the following situation for the aluminium rod described in Exercise 7.8. If a twisting torque, represented by a vector parallel to the rod axis, is applied to the end of the rod, then the rod will twist through an angle $\Delta\theta$. The twist is a measure of the shear strain produced by the torque-induced stress. In equilibrium, $\Delta\theta$ is determined by matching the size of the applied torque to that of the opposing torque involving the shear modulus of aluminium, which is 26×10^9 N/m^2. For small $\Delta\theta$ the shear stress–strain curve is linear, and the twist behaviour is reversible. If the mass is twisted through a small angle and is then released, it will perform oscillatory motion and the system constitutes a torsional oscillator as discussed in Chapter 8.

8 Oscillations

8.1 INTRODUCTION

Oscillatory behaviour is found in a wide variety of physical systems. Familiar examples include the to-and-fro motion of a child's playground swing, the periodic motion of the pendulum in a mechanical clock, and the oscillation of a mass suspended on a spring. For mechanical systems, this kind of motion can be described in detail using Newton's second law. The time-dependent motion is characterized by a frequency, which may, if the oscillation is only weakly damped, remain approximately constant over a long period.

The important case of simple harmonic motion is central to any discussion of oscillatory motion. Simple harmonic motion is explained in general terms in Section 8.2, and specific examples are given in Sections 8.3 to 8.5. Section 8.6 elaborates on the kinetic and potential energy contributions to the total energy of a mass that is undergoing oscillatory motion. The final sections of this chapter are concerned with, firstly, damped oscillators and, secondly, the response of an oscillator which is driven by an external mechanism at frequencies other than its natural frequency.

8.2 SIMPLE HARMONIC MOTION

Consider a system consisting of a mass m subject to a time-varying force $F(t)$, which causes the mass to undergo oscillatory motion parallel to the x-axis of a Cartesian coordinate system. Let $x(t)$ be the displacement of the mass from the origin at time t. The necessary condition for simple harmonic motion (SHM) to occur is that $x(t)$ be a periodic function of time of the form

$$x(t) = x_m \cos(\omega t + \phi) \tag{8.1}$$

with x_m the amplitude of the motion, ω the angular frequency, and ϕ the phase angle determined by the initial conditions. Note that the sine function could be used instead of the cosine function by changing the value of ϕ. Analogous to the case of circular motion in Chapter 5, the angular frequency is related to the period T for a complete

DOI: 10.1201/9781003485537-8

cycle of oscillatory motion by $\omega = 2\pi/T$. Newton's second law gives the time-dependent force acting on m as

$$F = m\,a = m\frac{d^2 x}{dt^2} = -m\,\omega^2\,x_m \cos\left(\omega\,t + \phi\right) = -m\,\omega^2\,x \tag{8.2}$$

Equation (8.2) shows that the force necessary to produce SHM of the mass m is proportional to the displacement, and, in view of the minus sign, acts towards the origin, that is in the opposite direction to the displacement. The classic example of a linear restoring force acting on a mass undergoing oscillatory motion is that of a mass attached to a spiral spring.

It is interesting and instructive to note the similarity of the expression describing SHM in 1D to the expressions for the x and y components in the rotating vector representation of circular motion in 2D given in Chapter 5. The angular displacement of a rotating vector, of length A, as a function of time is given by $\theta = \omega\,t + \phi$, and it follows that the amplitude of the x component is $x(t) = A \cos\left(\omega t + \phi\right)$, which is identical in form to Equation (8.1). The y component is analogous to this, but with the cosine function replaced by the sine function. Thus, circular motion can be viewed as a combination of two SHM motions along orthogonal axes with a phase difference $\Delta\phi = \pi/2$ between the motions. If the tip of the x component of the rotating vector were to be viewed in the plane of motion, looking down the y-axis, the 1D motion seen would correspond to SHM.

8.3 MASS ON A SPRING

8.3.1 HOOKE'S LAW

If a spiral spring is attached to a rigid support and is extended by a force applied to the free end, then, in equilibrium, the spring exerts an equal and opposite force to match the applied force. This scenario is illustrated in Figure 8.1 for a light spiral spring hung vertically with a mass m suspended from its free end. The length of the spring is extended by the applied force. Provided the extension Δl is not too large, the opposing force F exerted by the spring is found to obey Hooke's law, which states that $F \propto \Delta l$. Hooke's law is an empirical law, but it is found to hold remarkably well provided the extension does not produce permanent deformation of the spring. The proportionality constant used in Hooke's law is known as the spring constant k, giving the relationship

$$F = -k\,\Delta l \tag{8.3}$$

The SI units for k are N/m. The minus sign shows that the force acts in the opposite direction to the extension. Equation (8.3) is the required condition for SHM, as shown in Section 8.2.

Hooke's law follows from the elastic properties of solids as described in Chapter 7. The spring extensions discussed in this chapter are assumed to be in the elastic range, below the elastic limit.

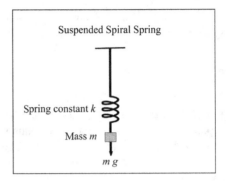

FIGURE 8.1   Depiction of a spiral spring suspended vertically with a mass m attached to the free end. In equilibrium, the spring exerts an upward force equal to the weight $m\,g$.

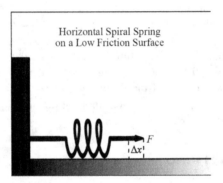

FIGURE 8.2   Representation of a spiral spring oriented horizontally parallel to x and subject to a force F which produces an extension Δx.

8.3.2   OSCILLATIONS OF A SPRING SYSTEM

To avoid having to consider gravitational effects when dealing with a mass on a spring, it is convenient to support the system on a horizontal surface, so that the downward gravitational force $m\,g$ on the attached mass m is matched by the upward reaction force produced by the surface. Coordinates are chosen with the horizontal x-axis along the longitudinal axis of the spring and the origin at the fixed end of the spring as illustrated in Figure 8.2. If the spring that is extended and released obeys Hooke's law, it follows that the mass will execute SHM provided the frictional damping forces are negligibly small. Comparing Equation (8.2) with Equation (8.3), it can be seen that $k = m\,\omega^2$ and the angular frequency of the SHM is therefore $\omega = \sqrt{k/m}$. This result for SHM can be obtained directly using Newton's second law together with Hooke's law as shown below.

In applying Newton's second law to the motion of the mass, and assuming that friction forces are negligible, it is necessary to consider just the horizontal force that is given by Hooke's law. Equation (8.1) gives the displacement of a mass undergoing SHM as $x(t) = A\cos(\omega t + \phi)$. The use of Newton's second law together with Hooke's law leads to the relationship

$$F = m\frac{d^2x}{dt^2} = -m\,\omega^2\,x_m\cos(\omega t + \phi) = -m\,\omega^2\,x = -k\,x \qquad (8.4)$$

The angular frequency is $\omega = \sqrt{k/m}$, as noted above, with oscillation period $T = \dfrac{2\pi}{\omega} = 2\pi\sqrt{\dfrac{m}{k}}$.

In the case of the spring–mass system being oriented vertically along the y-axis, the weight $m\,g$ extends the length of the spring by an amount Δy, which is given by Hooke's law as $\Delta y = \dfrac{m\,g}{k}$. In order to allow for this static extension due to the gravitational force, the origin O along y is shifted downwards by Δy. The spring constant k is unchanged and Hooke's law becomes $F = -k\,y$. Apart from the change of axis from x to y, Newton's second law retains the form used in the horizontal orientation case, with angular frequency again given by $\omega = \sqrt{k/m}$.

Exercise 8.1: A light spiral spring, with a mass $m = 0.05$ kg attached to one end, is located on a smooth horizontal surface with the other end clamped to a rigid support. If the oscillation period is 0.8 s, what is the spring constant? Determine the static extension of the spring when it is suspended vertically.

Using the result $T = 2\pi\sqrt{\dfrac{m}{k}}$ gives $k = \dfrac{4\pi^2\,m}{T^2} = \dfrac{39.5 \times 0.05}{2.56} = 3.08\,\text{N/m}$. When the spring is suspended vertically, with its cylindrical axis parallel to y, the static extension is given by Hooke's law as $\Delta y = \dfrac{m\,g}{k} = \dfrac{0.05 \times 9.82}{3.08} = 0.16$ m .

It is instructive to examine SHM for a spring oriented with its axis aligned vertically along the y-axis with a mass m attached to its free end as shown in Figure 8.1. The Earth's gravitational field exerts a *constant* downward force $F_G = m\,g$ on the mass leading to an extension of the spring. When set in vertical motion, the mass executes SHM about the extended length equilibrium position, taken as the origin. The Hooke's law force $F_H = -k\,y$ provided by extension and compression of the spring determines the motion of the mass. The frequency $\omega = \sqrt{k/m}$ of the SHM motion is precisely the same as that found when the spring axis lies along the x-axis. The gravitational force does play a role in determining the potential energy behaviour as a function of the position of the vertical oscillator, as is discussed in Section 8.6.

8.4 SIMPLE PENDULUM

The classic simple pendulum consists of a bob of mass m suspended on a string attached to a fixed support. The gravitational field exerts a force $F = m g$ on the bob and this keeps the string taut. When the bob is pulled to one side and then released, the system undergoes oscillatory motion about the vertical direction as depicted in Figure 8.3. In its motion, the bob sweeps out an arc of radius equal to the string length l.

Consider the forces acting on the bob when the string makes an angle θ with the vertical direction. The tension T in the cord acts perpendicular to the direction of motion of the bob as it swings along the arc to which its motion is constrained. The weight $m g$, which acts vertically downwards, can be resolved into two perpendicular components as follows. Firstly, $F_\perp = m g \cos \theta$ is parallel to the string and perpendicular to the direction of motion. Secondly, $F_\parallel = m g \sin \theta$ acts along the direction of motion as shown in Figure 8.3. It is convenient to describe the motion of the bob in terms of the arc length displacement $s = l \theta$ from the origin, which is chosen to lie at the bottom point of the arc of motion. Applying Newton's second law to the oscillatory motion gives

$$-m g \sin \theta = m \frac{d^2 s}{dt^2} = m l \frac{d^2 \theta}{dt^2} \qquad (8.5)$$

The minus sign is inserted because the force $F_\parallel$ acts in the opposite direction to the displacement. If the amplitude of the oscillating motion is sufficiently small, then to a good approximation $\sin \theta \simeq \theta$, and Equation (8.5) becomes

$$-\frac{g}{l} \theta = \frac{d^2 \theta}{dt^2} \qquad (8.6)$$

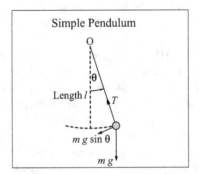

Simple Pendulum

FIGURE. 8.3   A simple pendulum consists of a bob of mass m suspended on a light string of length l. For small oscillation amplitude θ the pendulum executes motion approximating simple harmonic motion.

The comparison of Equation (8.6) with Equation (8.2) shows that for small θ the bob executes SHM, with the angular displacement at time t given by $\theta = \theta_m \cos(\omega t + \phi)$. Inserting this expression for θ into Equation (8.6), and simplifying, gives the angular frequency as

$$\omega = \sqrt{g/l} \qquad (8.7)$$

The period is $T = 2\pi\sqrt{\dfrac{l}{g}}$. The condition that $\sin\theta \simeq \theta$ holds to within 1% for angles

$\theta < \dfrac{\pi}{12}$ (or $\theta < 15°$). Equation (8.7) shows that the period of the pendulum does not depend on the mass of the bob, and it therefore applies universally for all pendulums of this type and length.

An alternative approach to the mechanics of the simple pendulum can be used if the pendulum string is replaced by a rigid rod of length l and negligible mass. The pendulum in this form is viewed as a rigid body, with moment of inertia about the suspension point given by $I = ml^2$, which is subject to a clockwise torque, of magnitude $\Gamma = mgl\sin\theta$, produced by the gravitational force acting on the bob as depicted in Figure 8.3. Newton's second law for rigid body motion is $\Gamma = I\alpha$, with $\alpha = \dfrac{d^2\theta}{dt^2}$, which gives $-mgl\sin\theta = ml^2\dfrac{d^2\theta}{dt^2}$. Taking $\sin\theta \simeq \theta$ for small θ, and simplifying, gives

$$-\frac{g}{l}\theta = \frac{d^2\theta}{dt^2} \qquad (8.8)$$

Equation (8.8) is identical to Equation (8.6), and the angular frequency is therefore given by $\omega = \sqrt{g/l}$ as before.

Exercise 8.2: A simple pendulum consists of a light rod, of length 0.8 m, with a mass m attached to the lower end. The pendulum is set in motion by displacing the bob by a small amount from its static equilibrium position. Determine the period of this pendulum. By how much would the period change if the length of the rod were halved?

Using the expression $T = 2\pi\sqrt{\dfrac{l}{g}}$ for the period gives $T = 2\pi\sqrt{\dfrac{0.8}{9.8}} = 1.8\,\text{s}$.

If the length of the rod is halved, the period would decrease by a factor $\sqrt{0.5} \approx 0.7$.

8.5   RIGID BODY OSCILLATIONS

If a rigid body is suspended on a pivot allowing rotational motion about an axis that does not pass through the centre of mass, it will execute SHM after being rotated through a small angle θ from its equilibrium orientation and then released. The situation is a generalization of the case of the simple pendulum consisting of a mass attached to one end of a light, but rigid, rod as discussed in Section 8.4.

Consider a solid slab of material chosen for convenience to be of rectangular shape, although the present analysis applies to arbitrarily shaped objects. The slab is suspended as shown in Figure 8.4, with the pivot axis at a distance L from the centre of mass. Using Newton's second law $\Gamma = I\,\alpha$ for the oscillatory motion of the body subject to a torque Γ, with I the moment of inertia of the body about the pivot axis and α the angular acceleration, gives, for small θ, the equation of motion as

$$-M\,g\,L\sin\theta \simeq -M\,g\,L\,\theta = I\,\frac{d^2\theta}{dt^2} \tag{8.9}$$

The magnitude of the torque produced by the gravitational force is given by $\Gamma = M\,g\,L\sin\theta \simeq M\,g\,L\,\theta$. As before, the approximation $\sin\theta \simeq \theta$ holds for small amplitude oscillations. The minus sign shows that the acceleration occurs in the opposite direction to the displacement.

Equation (8.9) can be written in the form $\dfrac{d^2\theta}{dt^2} = -\omega^2\theta$, with $\omega^2 = \dfrac{M\,g\,L}{I}$. This is the now familiar expression for SHM. Use of the parallel axis theorem for the moment of inertia of a rigid body about an axis at a distance L from a parallel axis through the centre of mass gives $I = I_{CM} + M\,L^2$, with I_{CM} the moment of inertia about the centre of mass. The angular frequency becomes $\omega = \sqrt{\dfrac{M\,g\,L}{I_{CM} + M\,L^2}}$. The moments of inertia of rigid bodies with various shapes about axes through their centres of mass are given

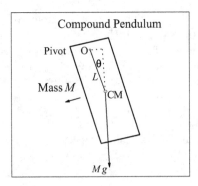

FIGURE 8.4   A rectangular-shaped rigid body of finite thickness and mass M is pivoted about an axis at O a distance L from its centre of mass. For small angular displacements θ the body executes SHM in the Earth's gravitation field.

in Chapter 6. Note that if the pivot axis passes through the centre of mass, then $L = 0$, and hence $\omega = 0$ or, equivalently, the period T goes to infinity.

Exercise 8.3: Determine the period of the oscillations of a solid disk of mass M and radius R about a perpendicular pivot axis near the edge of the disk.

Using the expression for the angular frequency of oscillation of a rigid body given by $\omega = \sqrt{\dfrac{M g L}{I_{CM} + M L^2}}$, together with the observations that $L = R$ and the moment of inertia about a perpendicular axis through the centre of mass is $I_{CM} = \dfrac{1}{2} M R^2$, leads to $\omega = \sqrt{\dfrac{M g R}{\dfrac{1}{2} M R^2 + M R^2}} = \sqrt{g / \dfrac{3}{2} R}$. The period $T = 2\pi \sqrt{\dfrac{3}{2} R / g}$ increases as the square root of the radius.

8.6   ENERGY OF OSCILLATION

A system consisting of a mass undergoing SHM has mechanical energy E made up of kinetic energy K and potential energy U. If damping forces are negligible, mechanical energy is conserved. Both K and U vary with time, but their sum E remains constant. If damping is not small, mechanical energy will not be conserved and the amplitude of the SHM oscillations will steadily decrease. For large damping, the system will not exhibit SHM. This case is discussed in Section 8.7.

As an example, consider a mass m attached to one end of a spiral spring with the other end of the spring clamped to a fixed support. As an idealization, the spring–mass system is assumed to be located on a frictionless horizontal surface. In addition, it is assumed that the mass of the spring is small compared to that of the attached mass m, and can therefore be neglected. If the spring is extended and then released, the mass will undergo SHM as described in Section 8.3. Note that it is the kinetic energy K of the *mass m* undergoing SHM that is considered, while the potential energy U is associated with extension or compression of the *spring* which provides the necessary time-varying force to move the mass. The displacement of the mass as a function of time is given by Equation (8.1) as $x(t) = x_m \cos(\omega t + \phi)$. The angular frequency is $\omega = \sqrt{k/m}$, with k the spring constant.

The maximum value of potential energy stored in the spring is determined from the work done by the applied force, $F(x) = k x$, producing the initial extension which starts the oscillatory motion. It follows that $U = W = \int_0^x k\, x \, dx = \dfrac{1}{2} k\, x^2$. When the mass is allowed to execute SHM, the potential energy oscillates in time between zero and $\dfrac{1}{2} k\, x_m^2$ as given by the following equation:

$$U = \frac{1}{2}k\,x_m^2\,\cos^2\left(\omega t + \phi\right) \tag{8.10}$$

In addition, Equation (8.10) shows that U has a parabolic dependence on x, and swings between the values 0, when $x = 0$, and $\frac{1}{2}k\,x_m^2$, when $x = x_m$.

The kinetic energy of the mass is given by $K = \frac{1}{2}m v_x^2 = \frac{1}{2}m\left(\dfrac{dx}{dt}\right)^2$, and so differentiating the expression $x(t) = x_m \cos\left(\omega t + \phi\right)$ gives

$$K = \frac{1}{2}m\,x_m^2\,\omega^2\,\sin^2\left(\omega t + \phi\right) = \frac{1}{2}k\,x_m^2\,\sin^2\left(\omega\,t + \phi\right) \tag{8.11}$$

K and U oscillate between the same energy limits, but with a phase difference of $\pi/2$ between them.

The total energy is given by

$$E = K + U = \frac{1}{2}k\,x_m^2\,\sin^2\left(\omega t + \phi\right) + \frac{1}{2}k\,x_m^2\,\cos^2\left(\omega t + \phi\right) = \frac{1}{2}k\,x_m^2 \tag{8.12}$$

The variations of K, U, and E with displacement x are shown in Figure 8.5.

The variations of E, K, and U with *time* are shown in Figure 8.6.

With the spring fully extended at $x = x_m$, the mass m is momentarily at rest with U at a maximum and K at a minimum. In contrast, K is at a maximum and U is at a minimum when $x = 0$. The average potential energy U_{av} is equal to the average kinetic

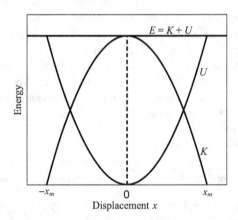

FIGURE 8.5 The kinetic energy K and the potential energy U as a function of displacement x for a mass m on a spiral spring undergoing simple harmonic motion. A minimum in K and a maximum in U occur when the displacement is $x = x_m$. For $x = 0$, the situation is reversed, with K a maximum and U a minimum. The total energy $E = K + U$ is constant for all x values.

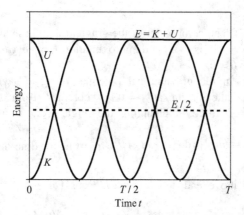

FIGURE 8.6 The potential energy U and the kinetic energy K as a function of time t for a simple harmonic oscillator. The total energy $E = U + K$ remains fixed, while U and K oscillate between 0 and E with a phase difference of π. $T = 2\pi/\omega$ is the period for a complete oscillation.

energy K_{av}, with $U_{av} = K_{av} = E/2$, as can be seen by considering the intersection of the curves in Figures 8.5 and 8.6.

The above discussion of the energy of a harmonic oscillator has considered a horizontally mounted spring with an attached mass moving on a frictionless surface. For a vertically mounted spring system, as shown in Figure 8.1, it is necessary to allow for the potential energy contribution of the attached mass as it undergoes vertical motion in the Earth's gravitational field. The two important forces acting on the mass when it is displaced from its static equilibrium position, which is chosen to be the origin of the vertical y-axis, are, firstly, $F_H = -k\,y$ due to the stretched spring and, secondly, $F_G = m\,g$ due to gravity. As shown above, the potential energy is obtained as a function of y by calculating the work done in stretching the spring. The calculation gives the relationship, $U = W = \int_0^y (k\,y + m\,g)\,dy = \frac{1}{2}k\,y^2 + m\,g\,y$. The maximum and minimum values of U are obtained with the appropriate choice of signs. Both U and K involve gravitational contributions and vary between zero and the altered upper and lower energy bounds as a function of time, in a similar way to the behaviour of the horizontal spring case shown in Figure 8.6. Since the gravitational *force* on mass m is constant, there is no change in ω as discussed in Section 8.3.

Exercise 8.4: Obtain expressions for the kinetic energy and the potential energy of a simple pendulum undergoing small angle oscillations approximating SHM.

From Section 8.4, the angular displacement of a simple pendulum bob suspended on a string of length l is $\theta = \theta_m \cos(\omega t + \phi)$, with $\omega = \sqrt{g/l}$ as given in Equation (8.7). The component of the gravitational force antiparallel to the displacement $ds = l\,d\theta$ is to a good approximation $F = -m\,g\,\theta$, where we have assumed that $\sin\theta \approx \theta$.

The increase in the gravitational potential energy U with angular displacement from the origin at $\theta = 0$ is obtained from the work done W by an applied force in changing the angle from 0 to θ. This gives

$$U = W = -F\,ds = m\,g\,l\int_0^\theta \theta\,d\,\theta = \frac{1}{2}m\,g\,l\,\theta^2 .$$ Inserting the time-dependent expres-

sion for θ given above leads to $U = \frac{1}{2}m\,g\,l\,\theta_m^2\cos^2\left(\omega t + \phi\right)$.

The kinetic energy is $K = \frac{1}{2}m\,v^2$, and using $v = l\dfrac{d\theta}{dt}$ gives $K = \frac{1}{2}m\,l^2\left(\dfrac{d\theta}{dt}\right)^2 =$

$\frac{1}{2}m\,l^2\,\omega^2\,\theta_m^2\sin^2\left(\omega t + \phi\right) = \frac{1}{2}m\,g\,l\,\theta_m^2\sin^2\left(\omega t + \phi\right)$

The total energy is constant and is given by $E = U + K = \frac{1}{2}m\,g\,l\,\theta_m^2$.

Note that in contrast to the horizontal spiral spring, where it is the elastic properties of the spring which determine the behaviour, in the case of a pendulum it is the Earth's gravitational field that gives rise to the potential energy properties associated with the motion. It is straightforward to consider the SHM behaviour of a vertically mounted spring carrying a mass m, which is subject to both elastic and gravitational forces that play a role in determining the energy of the system. This is left as an exercise.

Exercise 8.5: A thin strip of wood of length $2L$ and mass M is pivoted about an axis located at one end and oriented perpendicular to the strip's surface. To a good approximation, the strip executes SHM for small angular displacements θ from its static equilibrium orientation. Obtain expressions for the potential and kinetic energies of the strip as a function of time. Friction effects are negligible.

Newton's second law for rigid body oscillations, given in Equation (8.9), has

the SHM form $\dfrac{d^2\theta}{dt^2} = -\omega^2\,\theta$, with $\theta = \theta_m\cos\left(\omega t + \phi\right)$ where $\omega = \sqrt{\dfrac{M\,g\,L}{I}}$. L is

the distance from the pivot to the centre of mass of the object.

The kinetic energy is $K = \frac{1}{2}I\left(\dfrac{d\theta}{dt}\right)^2 = \frac{1}{2}I\,\omega^2\,\theta_m^2\sin^2\left(\omega t + \phi\right) = \frac{1}{2}M\,g$

$L\,\theta_m^2\sin^2\left(\omega t + \phi\right)$, while the potential energy is $U = M\,g\,L\int_0^\theta \theta\,d\theta = \frac{1}{2}M\,g\,L\,\theta^2 =$

$\frac{1}{2}M\,g\,L\,\theta_m^2\cos^2\left(\omega t + \phi\right)$.

The total energy $E = K + U = \frac{1}{2} M g L \theta_m^2$ is constant, as expected.

Note that the moment of inertia does not appear in the expression for K (due to the cancellation of terms), and the form of this expression is very similar to that obtained for the simple pendulum in Exercise 8.4, with the pivot to centre of mass distance L replacing the length of the simple pendulum string l. The compound pendulum expressions for U and K are widely applicable to SHM involving rigid bodies.

8.7 DAMPED SIMPLE HARMONIC MOTION

The discussion of SHM in this chapter has, until now, been concerned with ideal systems in which damping effects, due to friction or air resistance, are assumed to be negligibly small. In the limit of zero damping, SHM, once begun, persists indefinitely. In order to allow for damping it is necessary to introduce a damping force into Newton's second law. Consider a system comprised of a mass on a spring in the presence of a damping mechanism, which applies a velocity-dependent retarding force $F_D = -bv = -b\dfrac{dx}{dt}$ to the mass. This particular form for the retarding force applies, for example, if the damping is provided by a viscous liquid via a paddle attached to the mass. The equation of motion, which is based on Newton's second law, becomes

$$m\frac{d^2x}{dt^2} = -kx - b\frac{dx}{dt} \tag{8.13}$$

Although this second order differential equation is not straightforward to solve, considerable insight is gained by using the exponential form $x = x_m e^{-\beta t}$ as a trial solution of Equation (8.13), with β a parameter to be determined in terms of m, k, and b. Carrying out the differentiation of x and rearranging terms leads to the quadratic equation $\beta^2 - \dfrac{b}{m}\beta + \dfrac{k}{m} = 0$, with solution given by

$\beta = \dfrac{b}{2m} \pm \sqrt{\left(\dfrac{b}{2m}\right)^2 - \dfrac{k}{m}} = \dfrac{b}{2m} \pm \sqrt{\left(\dfrac{b}{2m}\right)^2 - \omega_0^2}$ where $\omega_0 = \sqrt{\dfrac{k}{m}}$ is the angular frequency of the undamped oscillator.

Three cases arise, referred to, respectively, as (1) underdamped with $\omega_0^2 > \left(\dfrac{b}{2m}\right)^2$, (2) critically damped with $\omega_0^2 = \left(\dfrac{b}{2m}\right)^2$, and (3) overdamped with $\omega_0^2 < \left(\dfrac{b}{2m}\right)^2$.

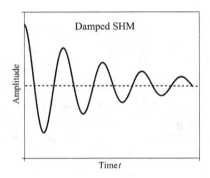

FIGURE 8.7   Decay in the amplitude of oscillations with time for a lightly damped simple harmonic oscillator returning to static equilibrium.

In the underdamped case, the solution involves the exponential of a complex function with real and imaginary parts, and can be shown to take the form

$$x = x_m \, e^{-t/\tau} \cos\left(\omega_D \, t + \phi\right) \tag{8.14}$$

with $\omega_D = \sqrt{\omega_0^2 - \left(\dfrac{b}{2m}\right)^2}$ the frequency of the decaying oscillations, and $\tau = 2m/b$ the decay constant. Figure 8.7 depicts the form of decaying oscillations for the underdamped case.

The critically damped case corresponds to the solution

$$x = x_m \left(1 + C\,t\right) e^{-t/\tau} \tag{8.15}$$

which shows that the amplitude of displacement of the mass decreases with no oscillatory behaviour. The return to the equilibrium position at $x = 0$ occurs most rapidly for critical damping. Finally, for overdamped conditions, the solution is a sum of two decreasing exponentials. Again, any initial displacement of the mass dies away with no oscillation, and experiment shows that the return to equilibrium becomes more sluggish as the damping increases. This behaviour corresponds to a decrease in β, which means that the solution for β with the negative sign ahead of the square root applies. The return to equilibrium over time of a spring-mass system following an initial displacement of the mass is shown in Figure 8.8 for slightly underdamped and critically damped cases.

From a transport perspective, whether by road or rail, critical damping is important. Motor car suspensions, for example, are designed to be critically damped so that following the traverse of a major obstacle or bump in the road, vehicles return to stability smoothly without oscillations, thus reducing the risk of discomfort or even motion sickness in passengers.

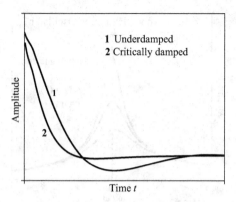

FIGURE 8.8   Gradual return to equilibrium over time for a simple harmonic oscillator which is (1) underdamped, and (2) critically damped. Overdamped behaviour (not shown) becomes more sluggish as the damping increases.

8.8 DRIVEN OSCILLATIONS

Consider a mechanical oscillator consisting of a mass m attached to a spring, with spring constant k, which executes SHM at a frequency $\omega_0 = \sqrt{k/m}$. Experiment shows that when a variable-frequency driving mechanism is used to apply an alternating force $F_A \cos(\omega t)$ to the mass, the system responds by executing oscillations at the driving frequency ω with an amplitude dependent on both F_A and ω. If the frequency is varied over a wide range, with F_A kept constant, the amplitude of oscillation reaches a maximum for $\omega = \omega_0$, a condition known as mechanical resonance. A familiar example of mechanical resonance is provided by a child's swing when it is pushed at its natural frequency.

For a mechanically driven horizontal spring plus mass oscillator, with allowance for damping, the equation of motion is obtained by adapting Equation (8.13) to give

$$m\frac{d^2 x}{dt^2} = -kx - b\frac{dx}{dt} + F_A \cos(\omega t) \tag{8.16}$$

The steady-state solution to Equation (8.16) has the familiar form $x = A(\omega) \cos(\omega t + \delta_\omega)$, but now both the amplitude and phase can be frequency dependent. The amplitude as a function of the driving frequency is found to be given by

$$A(\omega) = \frac{\dfrac{F_A}{m}}{\sqrt{\left(\omega^2 - \omega_0^2\right)^2 + \left(\dfrac{b}{m}\right)^2 \omega^2}} \tag{8.17}$$

This expression for the amplitude of oscillation as a function of the driving frequency is known as the damped oscillator form and is plotted in Figure 8.9 using $\omega_0 = 10 \text{ s}^{-1}$

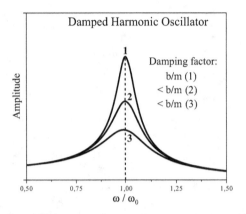

FIGURE 8.9 Amplitude response curves for a driven harmonic oscillator, obtained using Equation (8.17) and plotted versus the reduced frequency ω/ω_0 with ω_0 the resonance frequency of the undamped oscillator. As the damping factor b/m is increased the curves broaden and the maximum amplitude decreases.

and $F_A/m = 10$ N/kg. The damping constant, which is assumed to be much less than one ($b/m \ll 1$), determines the half-height width of the frequency response curve. Note that the maxima for the curves shown occur at a frequency slightly below the undamped harmonic oscillator resonance frequency.

The complete solution of Equation (8.16) for a driven harmonic oscillator provides information on the behaviour of the phase angle δ as a function of frequency and damping constant.

Exercise 8.6: A system consisting of a horizontally mounted spiral spring with a mass of 0.1 kg attached to one end has its oscillations damped by a velocity dependent retarding force $F_R = -b\,v$ with damping coefficient $b = 0.2$ kg/s. In the absence of external forces, the oscillation frequency is $\omega_0 = 30$ rad/s (i.e. the frequency is $f = 4.8$ Hz). If a variable-frequency oscillating driving force $F = F_A \cos(\omega t)$ with amplitude $F_A = 1.2$ N is applied to the mass parallel to the long axis of the spring, what will the amplitude of the oscillations be (a) for $\omega = \omega_0$ and (b) for $\omega_0 = 0.8\,\omega_0$?

Inserting $\omega = \omega_0$ together with the values for F_A and b in Equation (8.17) gives the amplitude of the driven oscillations as $A(\omega_0) = \dfrac{F_A}{b\,\omega_0} = \dfrac{1.2}{0.2 \times 30} = 0.2$ m .

Changing ω to $0.8\,\omega_0$ gives $A(\omega) = \dfrac{F_A/m}{\sqrt{\left(\omega^2 - \omega_0^2\right)^2 + \left(b/m\right)^2 \omega^2}} =$

$\dfrac{12}{\sqrt{\left(900 - 576\right)^2 + 4 \times 576}} = 0.037$ m.

As the damping constant is decreased, the response curve of a mechanical oscillator becomes more and more sharply peaked near the undamped centre frequency ω_0. The maximum amplitude response to the driving force then occurs for $\omega = \omega_0$. This is known as the resonance condition.

Mechanical resonance effects occur in a wide variety of physical situations, including musical instruments involving strings, air columns, and drumheads. Going beyond mechanical systems, resonance phenomena are also important in electromagnetism. For example, electronic-tuned circuits have numerous applications in communications and in specialized fields such as magnetic resonance imaging. Resonance effects are particularly important in molecular and atomic spectroscopy, and in related areas which include lasers and atomic clocks. While a detailed description of atomic-scale resonance effects requires familiarity with quantum mechanics, the ideas and results presented in this chapter, particularly those concerning damped and driven oscillators, are useful at length scales ranging from the macroscopic to the microscopic.

9 Waves in Low Dimensions

9.1 INTRODUCTION

Wave motion is encountered in a large variety of situations. Familiar examples include sound waves in the air and waves in the sea. There is an important distinction between mechanical waves, which require a medium in which to propagate, and electromagnetic waves, which propagate in a vacuum as discussed in Chapter 10. The subject is introduced by considering the properties of waves on a stretched string. The results obtained for waves on strings are useful in considering other types of waves, including sound waves and light waves. This chapter focuses on waves in a low number of dimensions, specifically with waves on strings and sound waves in pipes. Chapter 10 is concerned with waves in higher numbers of dimensions, in particular sound waves in air.

There are two types of mechanical waves, which are known as transverse waves and longitudinal waves, respectively. In order to illustrate the distinction between these wave types, consider a linear system consisting of a long chain of beads with neighbours connected together by springs. If a periodic force is applied at one end of the chain parallel to its long axis, then springs near that end experience a slight compression, followed by recovery to their original length, and then slight extension. With the passage of time, compression–expansion effects are transmitted along the chain as a longitudinal wave. In order to generate a transverse wave in the chain, the periodic force would be applied perpendicular to the chain axis, leading to periodic transverse motion involving spring compression–expansion effects, which propagate down the chain. For both wave types, energy is transmitted down the chain without any long-range displacement of the beads. By stacking the chains into 2D sheets or 3D blocks of coupled beads, it is possible to picture wave propagation in higher dimensional systems. As an example, wave motion in real solids can be simulated using a model in which the atoms or ions are coupled together by forces, which, for small displacements of the interacting particles, obey Hooke's law.

9.2 WAVES ON A STRING

9.2.1 TRAVELLING PULSES ON STRINGS

A long tautly stretched string allows transverse travelling waves to be generated on it by moving one end of the string up and down in a systematic way. A single pulse that travels along the string can be generated by a flick of the end, while a sinusoidal travelling wave can be produced using a mechanical oscillator attached to the

 DOI: 10.1201/9781003485537-9

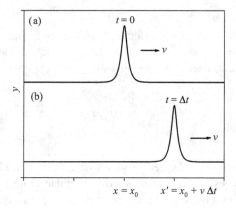

FIGURE 9.1 A pulse travelling with velocity v along a stretched string aligned parallel to the x-axis shown at times (a) $t = 0$ and (b) $t = \Delta t$. During time Δt the pulse travels a distance $\Delta x = v\,\Delta t$.

string. As a starting point, it is convenient to consider a single pulse propagating along an infinitely long string. In order to specify the time-dependent behaviour of the string due to a travelling pulse, a set of Cartesian coordinates are introduced with the x-axis parallel to the stretched string and the y-axis perpendicular to x and aligned along the direction of the transverse motion. Let the displacement y of the string as a function of position x and time t be given by $y = F(x,t)$ where F is a function yet to be determined. Figure 9.1 depicts the pulse location, firstly at time $t = 0$ and secondly at a later time $t = \Delta t$. For $t = 0$ let the peak amplitude of the pulse be y_0 at position x_0, while at the later time Δt the peak amplitude has propagated along the string and occurs at $x' = x_0 + v\,\Delta t$. Rearranging gives the relationship $x_0 = x' - v\,\Delta t$, which shows that the amplitude at point x' is equal to the amplitude at x_0 a time Δt earlier. Generalizing this approach to an arbitrary point x on the string leads to the form $y = F(x - vt)$ for the amplitude at that point at time t.

It is clearly of interest to consider the factors which determine the speed v of travelling waves on stretched strings. Observations and calculations given below show that two quantities are important, namely the string tension T and the mass per unit length μ. Other possible factors, such as the length of the string, are not important in determining v. It is possible to obtain an expression for v using dimensional analysis. This is done in Exercise 9.1.

Exercise 9.1: Use dimensional analysis to obtain an expression for the speed v of waves on a string in terms of the string tension T and the mass per unit length μ.

Let $v = T^\alpha \mu^\beta$ where α and β are exponents to be determined. The SI units of T are N, or $\mathrm{kg\,m\,s^{-2}}$, and those of μ are $\mathrm{kg\,m^{-1}}$. Inserting these units in the expression for v gives $\mathrm{m\,s^{-1}} = \left(\mathrm{kg\,m\,s^{-2}}\right)^\alpha \times \left(\mathrm{kg\,m^{-1}}\right)^\beta$. Grouping the exponents for "kg", "m", and "s" separately leads to the following simultaneous equations in α and β. For "kg", $0 = \alpha + \beta$; while for "m", $1 = \alpha - \beta$; and for "s", $-1 = -2\alpha$. Solving gives $\alpha = 1/2$ and $\beta = -1/2$. These exponent values lead to the following simple form for the speed of waves on a string:

$$v = \sqrt{\frac{T}{\mu}} \tag{9.1}$$

The above expression for the speed of waves on a string can also be derived using a model that simulates the motion of a string segment close to the maximum displacement position of a pulse travelling along the string. As a result of the travelling pulse, segments of the string exhibit curvature and are no longer straight. If a segment curves upwards, then the segment will experience a resultant force with an upward component along the y-direction due to asymmetry in the tension forces acting on the segment as illustrated in Figure 9.2. Adding vector components, the net force in the y-direction is $T\left(\sin\theta_2 - \sin\theta_1\right)$, where T is the tension in the string while θ_1 and θ_2 are the angles the tension force at the lower and upper ends of the segment make with the x-axis as given in Figure 9.2.

Next, the curved segment at the point of maximum displacement of the string in the y-direction can, as an approximation, be viewed as the arc of a circle of radius R, which is chosen to give the best fit in the vicinity of the peak. Let the segment of length l subtend an angle $\Delta\theta$ so that $l = R\,\Delta\theta$. Because of symmetry in the downward curvature, the tension forces at each end of the segment act downwards along $-y$, with each force making an angle $\Delta\theta/2$ with the x-direction as can be seen in Figure 9.3.

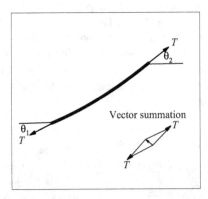

FIGURE 9.2   Tension forces acting on a string segment undergoing propagating wave motion. The inset shows the summation of the tension vectors T. The resultant has both vertical and horizontal components. Note that if a segment is straight, then $\theta_1 = \theta_2$ and the net force along y is zero.

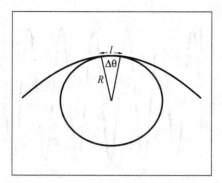

FIGURE 9.3 The shape of the string segment of length l through which the wave peak is moving can be well approximated by the arc of a circle of radius R, as shown, with $l = R\,\Delta\theta$.

If $\Delta\theta$ is chosen to be sufficiently small, then to a good approximation $\sin(\Delta\theta/2) \approx \Delta\theta/2$, and the net force on the segment is $F = 2T\sin(\Delta\theta/2) \approx T\,\Delta\theta$. Newton's second law then gives $F = T\,\Delta\theta = ma = \mu R\,\Delta\theta\,a$, where a is the acceleration of the segment in the $-y$ direction. The instantaneous acceleration of the segment is obtained using the expression $a = v^2/R$ for the centripetal acceleration of a mass in circular motion with speed v. Substituting for a in the Newton's law expression given above, and simplifying, gives $T = \mu v^2$. The resulting expression for the speed is $v = \sqrt{T/\mu}$, in agreement with the expression obtained by dimensional analysis in Exercise 9.1.

Exercise 9.2: A single pulse travels along a long string, of mass per unit length $\mu = 0.05$ kg/m, which is kept under tension $T = 4$ N. How long will it take the pulse to travel a distance of 3 m?
    The speed of the pulse is given by $v = \sqrt{T/\mu} = \sqrt{4/0.05} = 8.94$ m/s. The time taken to travel a distance of 3 m is $\Delta t = 3/8.94 = 0.34$ s.

9.2.2 HARMONIC WAVES ON STRINGS

Harmonic waves with wavelength λ are produced on a string by generating a periodic transverse displacement at one end of the string. This can be achieved by driving the transverse motion using a harmonic oscillator machine. The resultant waveform as a function of position along the string, in units of x/λ, shown at a fixed time $t = 0$, is a sine wave as given in Figure 9.4. The characteristic wavelength is the distance between wave crests. A similar plot is obtained for the vibrational amplitude at some fixed position as a function of time in units of t/T, with t the time and T the period for one transverse oscillation. Harmonic waves travel a distance λ in time T. If the speed of the wave along the string is v, then it follows that $v = \lambda/T = \lambda f$, where $f = 1/T$ is the frequency of the wave.

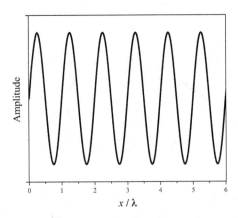

FIGURE 9.4 Harmonic wave on a string, whose amplitude as a function of x/λ is given by $y(x) = A\sin(2\pi x/\lambda)$ at a fixed time $t = 0$.

The wave function $F(x,t)$ for harmonic waves in 1D is obtained by adapting the general form $y = F(x - vt)$, given in Section 9.2.1 for the transverse y displacement of a string as a function of x and t, into the specialized form $y = y_m \sin k(x - vt)$, with y_m the amplitude (replacing A) and $k = 2\pi/\lambda$. Introduction of k converts the distance x along the string into an increasing angle in radians related to the wavelength by the variable x/λ as required. In addition, the term $2\pi vt/\lambda$ becomes ωt using the relationship $v/\lambda = f$ and the familiar relationship $2\pi f = \omega$. Thus, for harmonic waves the wave function is $F(x,t) = A\sin(kx - \omega t)$, with the displacement given by

$$y(x,t) = y_m \sin(kx - \omega t) \qquad (9.2)$$

Equation (9.2) is of central importance in developing a description of many wave-related phenomena, including the superposition of waves and standing waves.

Exercise 9.3: Harmonic travelling waves are generated on a long string using a 30 Hz mechanical oscillator. Taking the string tension as 3 N, and the mass per unit length as 0.06 kg/m, determine the wavelength.

From Equation (9.1), the wave velocity is given by $v = \sqrt{T/\mu} = \sqrt{3/0.06} = 7.1$ m/s. The relationship $v = \lambda f$ then gives $\lambda = 7.1/30 = 0.24$ m.

9.3   THE WAVE EQUATION

The wave equation for waves on strings, which is based on Newton's second law, is a second order partial differential equation with applications throughout wave physics.

In 1D, it will be shown that any wave function of the form $F(x - v\,t)$ involving the two variables x and t is a solution to the wave equation.

Consider a string of mass per unit length μ, under tension T, which is aligned along the x-axis of a Cartesian coordinate system. Travelling waves are generated on the string and result in displacements parallel to the y-axis. At some fixed time t, let a string element of length Δx be displaced through a distance y by the travelling wave. In Section 9.2, it is shown that the net force on a *segment* of a string that is undergoing a wave-induced displacement is, to a good approximation, given by $F = T\,\theta$, with $\theta = \theta_2 - \theta_1$ the difference in the angles that the tension T makes with x at the upper and lower ends of the segment. On a smaller scale, an *element* of the string experiences a net force in the y direction given by $F = T\left(\dfrac{\partial^2 y}{\partial x^2}\right)\Delta x$, where $\dfrac{\partial^2 y}{\partial x^2}$ is a measure of the string curvature provided the wave function is not a rapidly changing function of x. (The curvature gives the rate of change of the direction per unit length of the string.) Note that $\dfrac{\partial^2 y}{\partial x^2}$ may be positive, zero, or negative depending on whether the string curves upwards, is straight, or curves downwards. The use of partial derivatives is necessary because the variable t is being held constant. Taking the acceleration in the y-direction as $\dfrac{\partial^2 y}{\partial t^2}$, Newton's second law is written as $T\left(\dfrac{\partial^2 y}{\partial x^2}\right)\Delta x = \mu\,\Delta x\left(\dfrac{\partial^2 y}{\partial t^2}\right)$. Cancelling Δx leads to the wave equation for waves on a string,

$$\frac{\partial^2 y}{\partial x^2} = \frac{\mu}{T}\frac{\partial^2 y}{\partial t^2} \tag{9.3}$$

It has been shown previously that the speed of a wave on a string is $v = \sqrt{T/\mu}$, and substituting this expression in Equation (9.3) gives the relationship

$$\frac{\partial^2 y}{\partial x^2} = \frac{1}{v^2}\frac{\partial^2 y}{\partial t^2} \tag{9.4}$$

Equation (9.4) can be generalized to apply to sound waves in 2D or 3D.

Exercise 9.4: Show that for waves on strings the harmonic wave function $y = y_m \sin(kx - \omega t)$ is a solution to the wave equation.

For the chosen function, the evaluation of the two partial derivatives in Equation (9.4) gives $\dfrac{\partial^2 y}{\partial x^2} = \dfrac{\partial}{\partial x}\left(\dfrac{\partial y}{\partial x}\right) = -y_m\,k^2 \sin(k\,x - \omega t)$ and $\dfrac{\partial^2 y}{\partial t^2} = \dfrac{\partial}{\partial t}\left(\dfrac{\partial y}{\partial t}\right) = -y_m\,\omega^2 \sin(k\,x - \omega t)$. Substitution of these derivatives in Equation (9.4) leads to $-y_m k^2 \sin(kx - \omega t) = -\dfrac{1}{v^2} y_m \omega^2 \sin(kx - \omega t)$, which simplifies to give $v = \omega/k = \lambda f$ where use has been made of $k = 2\pi/\lambda$ and $\omega = 2\pi f$. This expression for the speed v agrees with the result obtained previously for travelling harmonic waves on strings. This proves that the harmonic wave function is a solution of the wave equation.

9.4   ENERGY OF HARMONIC WAVES ON STRINGS

Travelling waves on a string transport mechanical energy away from the oscillator source that produces the waves. This energy flow results in time-dependent kinetic and potential energies being associated with each string segment. The kinetic energy of a segment Δl involves its *transverse* speed $v_y = \dfrac{\partial y}{\partial t}$, while the potential energy involves the slope of the segment $\dfrac{\partial y}{\partial x}$. The use of the harmonic wave function in Equation (9.2) gives $\dfrac{\partial y}{\partial t} = -y_m \omega \cos(kx - \omega t)$ for v_y and $\dfrac{\partial y}{\partial x} = y_m k \cos(kx - \omega t)$ for the slope. Note the use of partial derivatives in which one of the two independent variables in the wave function, either x or t, is allowed to vary, while the other is kept constant.

As before, the string tension is taken as T and the mass per unit length as μ. The kinetic energy of the string segment, $K = \dfrac{1}{2}\Delta m \, v_y^2$, can be written down immediately as

$$K = \frac{1}{2}\mu \, \Delta l \left(\frac{\partial y}{\partial t}\right)^2 = \frac{1}{2}\mu \, \Delta l \, \omega^2 \, y_m^2 \cos^2(kx - \omega t) = K_{max} \cos^2(kx - \omega t) \quad (9.5)$$

Equation (9.5) shows that K varies between zero and $K_{max} = \dfrac{1}{2}\mu \, \Delta l \, \omega^2 \, y_m^2$, reaching its maximum value *twice* in each period T as illustrated in the plot of K/K_{max} versus t/T in Figure 9.5. The plot is for a string element at a fixed position x in the string. Note that K_{max} depends on the mass of the string element multiplied by the product of the frequency squared and the wave amplitude squared.

The potential energy is obtained by calculating the work done by the propagating wave when it stretches a string segment further. If the string is in equilibrium, then $\dfrac{\partial y}{\partial x} = 0$ and the potential energy is zero. In the presence of a wave, the length of a string segment increases from Δx to Δl, and the work done is

$$U = T(\Delta l - \Delta x) = T\left(\sqrt{(\Delta x)^2 + (\Delta y)^2} - \Delta x\right) = T \, \Delta x \left(\sqrt{1 + \left(\frac{\partial y}{\partial x}\right)^2} - 1\right). \quad \text{In order to}$$

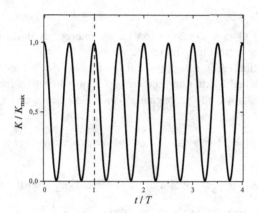

FIGURE 9.5 The kinetic energy ratio K/K_{max} for a string element as a function of time plotted in units of the period T of the harmonic waves which propagate along the string. The plot is based on Equation (9.5) with $x = 0$ to avoid a phase shift at $t = 0$.

simplify the calculation, it is assumed that the wave's amplitude is small and its wavelength λ is long compared to the length Δl of the segment. This implies that $\dfrac{\partial y}{\partial x} \ll 1$'

and therefore the square root term can be expanded using $\sqrt{1 + \left(\dfrac{\partial y}{\partial x}\right)^2} \approx 1 + \dfrac{1}{2}\left(\dfrac{\partial y}{\partial x}\right)^2$,

which leads to $U = \dfrac{1}{2}T\,\Delta x\left(\dfrac{\partial y}{\partial x}\right)^2$. The potential energy expression is obtained by

using $T = \mu v^2 = \mu\left(\dfrac{\omega^2}{k^2}\right)$, which gives, finally,

$$U = \frac{1}{2}T\,\Delta x\left(\frac{\partial y}{\partial x}\right)^2 = \frac{1}{2}\mu\,\Delta x\,\omega^2\,y_m^2\cos^2\left(kx - \omega t\right) = U_{max}\cos^2\left(kx - \omega t\right) \quad (9.6)$$

Comparing Equation (9.5) (after replacing Δl by Δx in the expression for K_{max}) with Equation (9.6), it is seen that $K = U$ with maxima and minima coinciding. The plot for U as a function of time is the same as the plot of K versus t given in Figure 9.5.

The total energy E of the string segment is given by

$$E = K + U = \mu\,\Delta x\,\omega^2\,y_m^2\cos^2\left(kx - \omega t\right) \quad (9.7)$$

The average transmitted energy E_{av} is obtained using the value $\dfrac{1}{2}$ for the average of

$\cos^2\left(kx - \omega t\right)$ over a cycle. This gives $E_{av} = \dfrac{1}{2}\mu\,\Delta x\,\omega^2\,y_m^2$.

Exercise 9.5: Obtain an expression for the energy per unit time transmitted along a string by a travelling harmonic wave.

From Equation (9.7), it follows that the rate at which energy passes along a string is obtained by dividing the energy E of a string segment by the time interval Δt for the wave to propagate a distance Δx and is given by

$$\frac{E}{\Delta t} = \mu \left(\frac{\Delta x}{\Delta t} \right) \omega^2 \, y_m^2 \cos^2 \left(k\,x - \omega t \right) = \mu\,v\,\omega^2\,y_m^2 \cos^2 \left(k\,x - \omega t \right)$$

Averaging $\cos^2 \left(k\,x - \omega t \right)$ over a cycle at a fixed point on the string gives the average power propagating along the string as $P_{av} = \dfrac{1}{2}\mu\,v\,\omega^2\,y_m^2$.

9.5 WAVE INTERFERENCE ON STRINGS

It is possible to generate two or more travelling waves on strings held under tension. The waves may travel in the same direction or in opposite directions, and can have a variety of forms from single pulses to harmonic waves. In examining how the waves interfere and give rise to modified wave functions in regions in which the primary waves overlap, it is necessary to make use of an established principle called the principle of superposition. The principle of superposition for interfering waves states that the amplitude of the resultant wave is given by the algebraic sum of the separate wave amplitudes. Symbolically the principle of superposition is expressed as

$$y(x,t) = \sum_i y_i (x,t) \tag{9.8}$$

where $y(x,t)$ is the amplitude of the resultant wave produced by the i interfering waves.

As a simple example, consider the interference of two harmonic waves of the same amplitude y_m and angular frequency ω, but with a phase difference ϕ between them. The waves travel in the $+x$ direction. (Note that if the waves have the same frequency and speed, then they have the same wavelength.) Using the superposition principle, the wave function of the resultant wave is given by $y_R (x,t) = y_m \sin(k\,x - \omega t + \phi) + y_m \sin(k\,x - \omega t)$. It is possible to carry out the summation using the trigonometric identity $\sin\alpha + \sin\beta = 2\sin\dfrac{\alpha+\beta}{2}\cos\dfrac{\alpha-\beta}{2}$. This gives

$$y_R (x,t) = 2 y_m \sin\left(k\,x - \omega t + \frac{1}{2}\phi \right)\cos\left(\frac{1}{2}\phi \right) \tag{9.9}$$

The amplitude of the resultant wave function is $2 y_m \cos(\phi/2)$. It follows that the amplitude will vary between $2 y_m$ and $-2 y_m$ as ϕ takes values from 0 to 2π, with total cancellation occurring for $\phi = \pi$ and higher odd multiples of π. The terms constructive and destructive interference are used, respectively, to describe the amplitude enhancement and amplitude reduction of the resultant wave compared to the amplitudes of the two interfering waves. Figure 9.6 shows two waves of the same amplitude and wavelength with $\phi = \pi$. Destructive interference leads to zero resultant amplitude for the superposition of these waves.

If the phase difference is taken as $\phi = \pi/2$ instead of π, as used in Figure 9.6, then Equation (9.9) predicts a resultant harmonic wave with amplitude $2y_m \cos(\pi/4) = 1.414\, y_m$ and phase angle $\phi = \pi/4$. Numerical calculations for this case, with $y_m = 1$, are shown in Figure 9.7. The resultant wave matches the predictions.

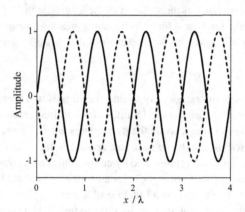

FIGURE 9.6 Two harmonic waves, shown by the solid and dash lines respectively, have the same amplitude and wavelength with a phase difference $\phi = \pi$. Wave crests coincide with troughs, and therefore superposition of the waves results in total destructive interference.

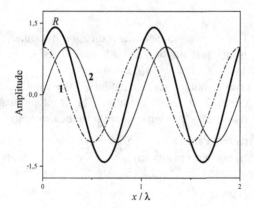

FIGURE 9.7 Two harmonic waves, labelled 1 and 2, have equal amplitudes $y_m = 1$ and wavelengths λ, with a phase difference of $\pi/2$. The resultant wave R, obtained by numerical superposition of waves 1 and 2, has amplitude 1.414 and phase angle $\pi/4$ as predicted by Equation (9.9).

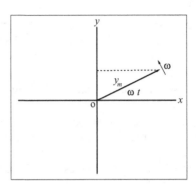

FIGURE 9.8 Representation of the variation of the amplitude $y(x,t)$ of a harmonic wave for given position x as a function of time t by a rotating phasor with angular frequency ω. The time-varying amplitude is given by the projection of the phasor length component onto the y-axis.

The use of a trigonometric identity to find the sum of two harmonic waves works well only for the special case of equal amplitudes and frequencies of the two waves. If the amplitudes are not equal, while the frequencies are the same, then a geometrical approach can be used to find the sum. In this approach, the amplitudes and phases of the two interfering waves are represented by counterclockwise rotating vectors, called phasors, with angular velocity ω in a Cartesian frame as indicated in Figure 9.8.

The amplitude of each wave as a function of x and t is given by the y-component of the corresponding rotating phasor in this geometrical representation of wave motion. In carrying out the vector summation, the orientations of the vectors are frozen corresponding to the situation at a particular position x at time t with due regard to any phase difference ϕ between the waves. An illustrative example is given in Exercise 9.6.

Exercise 9.6: Find the amplitude of the resultant of two interfering transverse waves 1 and 2, of unequal amplitudes, with wave functions given by $y_1(x, t) = y_1 \sin(k\,x - \omega\,t)$ and $y_2(x, t) = y_2 \sin(k\,x - \omega\,t + \phi)$.

Figure 9.9 gives a phasor representation of the amplitudes of the interfering waves 1 and 2 at fixed x and t. The resultant wave function is $y_R(x,t) = y_R \sin\left(k\,x - \omega t + \phi_R\right)$ with y_R and ϕ_R to be determined.

To simplify matters x and t have been chosen so that $y_1(x,t) = 0$, with phasor 1 parallel to the x-axis. To avoid confusion with the y-components of the phasors, the lengths are denoted by l_1 and l_2. Noting that $\theta = \phi$, the amplitude of the resultant wave R is obtained using the cosine rule $l_R = \sqrt{l_1^2 + l_2^2 - 2l_1l_2 \cos\left(\pi - \phi\right)}$, while the phase angle ϕ_R is given by $\tan\phi_R = \dfrac{l_2 \sin\phi}{l_1 + l_2 \cos\phi}$.

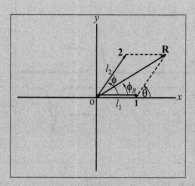

FIGURE 9.9 Addition of two harmonic waves with different amplitudes using a phasor diagram. The y-component of the resultant phasor R is equal to that of phasor 2, because the y-component of phasor 1 has been set to be zero, for convenience, at the time chosen for the phasor diagram representation. The resultant wave amplitude as a function of time is given by the y-component of R when rotating with angular frequency ω.

The solutions for the resultant amplitude and phase angle apply in general to two interfering waves of equal or unequal amplitudes. For waves of equal amplitude, symmetry considerations show that $\phi_R = \phi/2$, with the wave function taking the form given in Equation (9.9).

The wave interference effects described above involve waves with the same angular frequency ω. For harmonic waves with similar but different frequencies, interference effects give rise to what are known as beats as described in Chapter 10.

9.6 STANDING WAVES ON STRINGS

For waves propagating on a string of finite length, experiment shows that the boundary conditions at the ends of the string play a crucial role in determining how waves propagate on the string. Under certain conditions, as discussed below, standing waves are established. This development occurs when resonance conditions are achieved, with the string length an exact multiple of the wavelength, or of some fraction of the wavelength, for harmonic waves. String instruments make use of these standing wave effects in producing music.

Consider a wave propagating on a string, which reaches a boundary at which the wave is reflected back along the string. If the boundary is a rigid fixed support to which the string is attached, the reflected wave undergoes a phase change given by $\Delta\phi = \pi$. This change in phase can be understood using Newton's third law concerning action and reaction forces. The incoming wave causes the string to exert a force on the rigid support, which in turn exerts an equal and opposite reaction force on the string. The reflection process is illustrated in Figure 9.10.

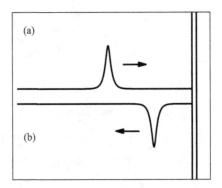

FIGURE 9.10   A transverse pulse travelling along a stretched string is reflected at a fixed boundary. The pulse is reflected with a phase change $\Delta\phi = \pi$, corresponding to inversion of the pulse amplitude.

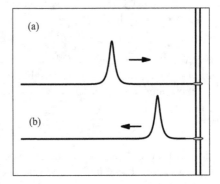

FIGURE 9.11   Reflection of a travelling pulse on a stretched string by a flexible boundary, which consists of a ring that is free to move with negligible friction along a fixed rod. The pulse is reflected with no change of phase, that is,, $\Delta\phi = 0$.

If instead of being fixed the boundary is flexible, the situation changes and there is no phase change in the reflection process so that $\Delta\phi = 0$. A flexible boundary can be achieved by attaching the string to a ring that can slide on a fixed rod with negligible friction. The ring moves in response to the force produced by the string due to the incoming wave as shown in Figure 9.11. In this case, the reflected wave undergoes no phase change and $\Delta\phi = 0$.

Generalizing the above results for travelling pulses on strings to all waves on strings, including harmonic waves, leads to the following reflection conditions: waves on a string that undergo reflection at a boundary experience phase changes $\Delta\phi = \pi$ at a fixed boundary, and $\Delta\phi = 0$ at a flexible boundary.

Following reflection at a boundary, a wave on a string will travel back along the string. If it meets another wave, interference will occur as the waves pass each other. The present discussion focuses on harmonic waves produced by a mechanical

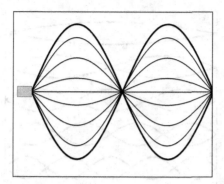

FIGURE 9.12   Representation of standing waves of wavelength λ on a string. The grey box depicts a mechanical wave generator, which produces harmonic waves that are reflected back and forth at the boundaries, leading to a buildup in the amplitude of oscillations. The bold lines represent the maximum amplitude of string oscillations, while fainter lines represent amplitudes at various times in an oscillation cycle.

oscillator to which the string is attached at one end. The string is kept under tension by fastening it to a support, which is either fixed or flexible. Experiment shows that when the conditions for standing waves are met, the amplitude of standing wave maxima is large compared to the amplitude of the mechanical oscillator vibrations. As illustrated in Figure 9.12, standing wave patterns are characterized by successive maxima and minima of vibrational amplitudes along the string. The maxima occur at positions called antinodes and the minima at nodes. The standing wave conditions in terms of string length in wavelength units are different for the fixed and flexible boundary cases. The fixed boundary case is dealt with first.

   Consider a taut string of length L which is fixed at one end to a source of low amplitude harmonic waves and at the other end to a rigid support as shown in Figure 9.12 for the particular case $L = \lambda$. A harmonic wave from the source travels along the string until it reaches the fixed support where it undergoes a phase change $\Delta\phi_1 = \pi$ and is reflected back towards the source. On reaching the source, which is treated as a rigid support, the wave is again reflected with a further phase change $\Delta\phi_2 = \pi$. In addition, the phase of the wave after reflection depends on x, the distance travelled as the wave makes the round trip from the source to the rigid support and back. If $x = n\,\lambda$, where n takes integer values $1, 2, 3, \ldots$, then $\Delta\phi_3 = 2\pi n$. When the combined phase shifts (i.e. $\Delta\phi_1 + \Delta\phi_2 + \Delta\phi_3$) are a multiple of 2π, the reflected wave is synchronized with the wave source. The wave amplitudes add up leading to large amplitude standing wave antinodes following multiple reflections. The distance travelled by a wave in a round trip along the string is $x = 2L$, and it follows that the condition for standing waves is given by

$$L = n\frac{\lambda_n}{2}, \text{ or } \lambda_n = \frac{2L}{n} \qquad (9.10)$$

with $n = 1, 2, 3, \ldots$. This gives L as a multiple of half wavelengths $\lambda_n/2$. A particular standing wave frequency f is obtained in terms of the string length using $f_n = v/\lambda_n$.

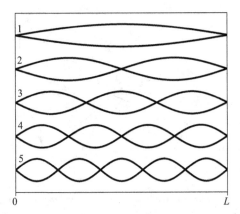

FIGURE 9.13   Standing waves on a string of length L with fixed boundaries at both ends. The integer values shown are the number of half wavelengths n in each mode as given by Equation (9.10).

This gives $f_n = \dfrac{n\,v}{2L}$, with v the speed of waves on the string. The various standing wave modes are known as harmonics, with the first harmonic corresponding to $n = 1$, the second harmonic to $n = 2$, and so on. The standing wave patterns which are given by Equation (9.10) are shown in Figure 9.13.

The case of a taut string attached to a flexible support at one end and a small amplitude harmonic wave source (which effectively provides a fixed support) at the other end is similar to that of the string attached to rigid supports at both ends as discussed above. A major difference is that no phase change occurs in the wave reflection process at the flexible end, so that $\Delta\phi_1 = 0$ there. Back at the wave source, which is regarded as a fixed end because of the small oscillation amplitude, reflection still occurs with a phase change $\Delta\phi_2 = \pi$. For the reflected wave to be synchronized with the wave source, following reflection, the total phase change in a complete round trip must be 2π or a multiple thereof. It follows that the phase change associated just with wave travel from source to flexible support and back should be an odd multiple of π given by $\Delta\phi_3 = n\,\pi$, with $n = 1, 3, 5, \ldots$. The standing wave condition in terms of L is

$$L = n\frac{\lambda_n}{4} \qquad (9.11)$$

with n an odd integer as given above. Standing waves thus occur when L is an odd integer multiple of a quarter wavelength. The standing wave frequencies are given by $f_n = \dfrac{n\,v}{4L}$. Harmonics in this case are called the first harmonic for $n = 1$, the third harmonic for $n = 3$, and on to the higher harmonics. Figure 9.14 shows the standing wave patterns up to the ninth harmonic.

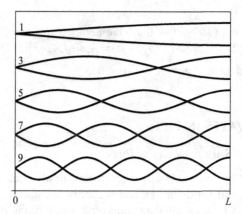

FIGURE 9.14 Standing wave harmonic patterns are shown for a string of length L with a fixed boundary at one end and a flexible boundary at the other. The integer values are the number of quarter wavelengths in each mode as given by Equation (9.11).

A harmonic wave source at frequency f_n can be used to generate the nth harmonic standing wave on a stretched string. String elements execute SHM parallel to the y-axis with frequency ω_n and amplitude dependent on the position x of the element along the string. For the nth harmonic, the standing wave amplitude at position x is given by $y_n(x) = A_n \sin(k_n x)$, with $k_n = 2\pi/\lambda_n$ and A_n the amplitude maximum value for the nth harmonic. As stated above, the amplitude of string vibration is a maximum at an antinode and a minimum at a node. The wave function for the nth harmonic therefore has the form

$$y_n(x,t) = A_n \sin(k_n x)\cos(\omega_n t + \varphi) \qquad (9.12)$$

with the phase angle φ determined by the initial conditions. The amplitude A_n initially increases with time and then stabilizes when the energy loss per cycle is equal to the energy input to the vibrating string from the harmonic wave source.

Exercise 9.7: A string of length 1.8 m and mass per unit length 50 g/m is attached to effectively rigid supports at both ends. (As already mentioned above, the source of harmonic waves is regarded as a rigid boundary.) The string is kept at a tension of 18 N. Find the frequency of the third harmonic. Give the form of the wave function for this harmonic.
The speed of waves on the string is $v = \sqrt{T/\mu} = \sqrt{18/0.05} = 19$ m/s. From Equation (9.10), the wavelength of the third harmonic is $\lambda_3 = 2L/3 = 2 \times 1.8/3 = 1.2$ m. Using $v = f\lambda$ gives the frequency of the third harmonic as $f_3 = 19/1.2 = 15.8$ Hz.

Using Equation (9.12), the third harmonic wave function is given by $y_3(x,t) = A_3 \sin(2\pi x/\lambda_3)\cos(2\pi f_3 t + \varphi) = A_3 \sin(5.24x)\cos(99.3t + \varphi)$. The amplitude A_3 and the phase φ are not specified.

9.7 SOUND WAVES IN PIPES

9.7.1 HARMONIC SOUND WAVES IN PIPES

Sound waves are different to 1D waves on strings, firstly because they can propagate in both fluids and solids in 3D, and secondly because they involve longitudinal pressure variations rather than transverse oscillations of string segments. There are, however, underlying common features of quasi-1D sound waves in pipes and waves on strings that lead to similarities in the physical descriptions of these wave phenomena. These similarities are brought out most clearly if a longitudinal displacement of mass elements description is used for sound waves in pipes. The present discussion focuses on sound waves in pipes containing a gaseous medium such as air. Both the pressure wave and mass element displacement descriptions are introduced.

As a starting point, harmonic sound waves travel along a pipe which encloses a gas that transmits the sound. The harmonic waves can be generated using a vibrating diaphragm as a source of plane waves of frequency f, which travel parallel to the x direction along the axis of the pipe. Figure 9.15 illustrates the situation, showing a cross-section of a pipe down which pressure waves are transmitted with a harmonic source at one end. Also shown is a representative disk-shaped gas element through which the waves pass.

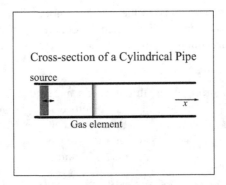

FIGURE 9.15 Transmission of harmonic pressure waves down a cylindrical pipe aligned along the x-axis. A cross-sectional view of the pipe is shown, with a harmonic longitudinal wave source at one end. Pressure fluctuations occur in disk-shaped elements as the waves travel down the pipe.

By analogy to harmonic waves on strings, the wave function for the harmonic sound waves is expressed in terms of gas pressure fluctuations $\Delta P(x,t)$ in disk-shaped regions oriented perpendicular to the tube axis, and has the same form as that given in Equation (9.2):

$$\Delta P(x,t) = \Delta P_m \sin(kx - \omega t) \tag{9.13}$$

The amplitude of the sound wave, ΔP_m, is given by the maximum pressure excursion above ambient pressure that is experienced in a disk region, while $k = 2\pi/\lambda$ and $\omega = 2\pi f$ as before. Note that by considering a planar sound wave travelling along a pipe, the 3D features of sound waves travelling in free space have been suppressed, and the spatial description is reduced to quasi-1D. The results that are obtained are generalized to 3D sound waves in Chapter 10.

An alternative form of the harmonic wave function, involving the displacement $d(x,t)$ of a mass element parallel to the x-axis, is given by

$$d(x,t) = x_m \sin\left(kx - \omega t - \frac{\pi}{2}\right) \tag{9.14}$$

where x_m is the amplitude of the oscillatory motion. It is important to note that the wave function expressed in terms of pressure variations is $\pi/2$ out of phase with the displacement wave function. This phase difference can be understood by appreciating that when the displacement of a fluid element along x is a maximum in its periodic motion, the pressure difference ΔP between the two sides of the fluid element is momentarily zero. Comparison of Equations (9.13) and (9.14) shows that ΔP_m is proportional to x_m, and dimensional analysis gives $\Delta P_m = v \rho \omega x_m$, with v the speed of sound in the medium.

The similarity of the wave functions for harmonic waves on strings, as given in Equation (9.2), and 1D harmonic sound waves in pipes is clear. This feature in pipes provides the basis for unifying the discussion of these diverse types of 1D waves. For example, it is simple to adapt the expressions for standing waves on strings to standing sound waves in open or closed pipes as is shown below. These results are particularly useful in considering the design and operation of musical instruments.

An expression for the speed of sound waves in fluids can be obtained by applying Newton's second law to the dynamics of a disk-shaped element in a fluid contained in a pipe of cross-sectional area A. The element, which is of length Δx, has volume $V = A\,\Delta x$ and is traversed in a time Δt by a pressure pulse travelling through the fluid at the velocity of sound v. If the pressure pulse has amplitude ΔP, the pressure difference across the length of the element during the time Δt leads to a small compression of the element and a slight change in its velocity δv in the x direction along which the pulse travels. It is convenient to express the fractional change in the volume of the element $\delta V/V$ in terms of the bulk modulus B of the fluid using the expression $\Delta P = -B \dfrac{\delta V}{V}$ from Chapter 7. The net *force* on the element during Δt is

$F = A\,\Delta P = -A\,B\dfrac{\delta V}{V}$, while the *rate of change of momentum* is given by $\dfrac{\Delta p}{\Delta t} = m\dfrac{\delta v}{\Delta t}$.

Note that the volume of the element decreases slightly during passage of the pulse, and therefore $\delta V = A\,\delta x < 0$. Writing the mass of the element in terms of the fluid density ρ as $m = \rho\,A\,\Delta x$, and then inserting the expressions for the force on the element, and for the rate of change of its momentum, into Newton's second law, and lastly cancelling A gives

$$-B\frac{\delta V}{V} = \rho\,\Delta x\,\frac{\delta v}{\Delta t} = \rho\,v\,\delta v = \rho\,v^2\,\frac{\delta v}{v} \tag{9.15}$$

The factor $\dfrac{\delta v}{v}$ can be related to the fractional volume change as follows:

$$\frac{\delta v}{v} = -\left(\frac{\delta x}{\Delta t}\right) \Big/ \left(\frac{\Delta x}{\Delta t}\right) = -\frac{A\,\delta x}{A\,\Delta x} = -\frac{\delta V}{V}.$$

Using this result in Equation (9.15) gives

$$v = \sqrt{\frac{B}{\rho}} \tag{9.16}$$

This simple expression for the speed of sound in a fluid has a similar form to that of a wave on a stretched string, $v = \sqrt{T/\mu}$ as given in Equation (9.1) with T the tension and μ the mass per unit length of the string.

Exercise 9.8: Obtain values for the speed of sound in air and water at ambient temperature using the values for their density ρ and bulk modulus B given below.

Water: $\rho_w = 1.0 \times 10^3$ kg/m^3 and $B_w = 2.2$ GPa
Air: $\rho_a = 1.21$ kg/m^3 and $B_a = 142$ kPa

Inserting the values for B and ρ in Equation (9.16) gives the speed of sound in water as $v_w = \sqrt{2.2\times10^9/1.0\times10^3} = 1480$ m/s. Similarly, for air the speed is $v_a = \sqrt{1.42\times10^5/1.21} = 343$ m/s. The speed of sound in water is approximately four times higher than the speed in air.

9.7.2 ENERGY OF HARMONIC SOUND WAVES IN PIPES

Making use of the wave function in Equation (9.14), although omitting its phase offset of $\pi/2$, the kinetic energy associated with the oscillatory motion of a fluid element with volume $\Delta V = A\,dx$ is given by

$$K = \frac{1}{2}m\,v^2 = \frac{1}{2}\rho\,\Delta V\left(\frac{\partial}{\partial t}d(x,t)\right)^2 = \frac{1}{2}\rho\,\Delta V\,\omega^2\,x_m^2\cos^2(kx-\omega t) \quad (9.17)$$

where ρ is the density of the fluid and A is the cross-sectional area of the pipe. Note the similarity of Equation (9.17) to Equation (9.5) for harmonic waves on strings.

Guided by the result obtained previously for string segments, the potential energy of a fluid segment traversed by sound waves is taken to be equal to the kinetic energy, and is thus given by

$$U = \frac{1}{2}\rho\,\Delta V\,\omega^2\,x_m^2\cos^2(kx-\omega t) \quad (9.18)$$

The total energy of the wave segment as a function of x and t follows as

$$E = K+U = \rho\,\Delta V\,\omega^2\,x_m^2\cos^2(kx-\omega t) \quad (9.19)$$

Averaging over a complete wave cycle gives the average energy as

$$E_{av} = \frac{1}{2}\rho\,\Delta V\,\omega^2\,x_m^2 \quad (9.20)$$

Using $\Delta V = A\,\Delta x$, it follows from Equation (9.20) that the average rate of energy transmission along the pipe is $E_{av}/\Delta t = \frac{1}{2}\rho\,A\,v\,\omega^2\,x_m^2$.

9.7.3 STANDING SOUND WAVES IN PIPES

Harmonic sound waves in pipes can be described using either of the wave function forms for sound waves that are given in Equations (9.13) and (9.14). It is convenient to use Equation (9.14) as a starting point, because of the similarity of the analysis to that given in Section 9.6 for standing waves on strings. Consider standing waves that are produced in a pipe that is open at one end and closed at the other. The displacement description requires a node at the closed end and an antinode at the open end. However, in making a comparison with standing waves on strings, there are notable differences in the reflection processes. Firstly, for sound waves there is no 180° phase change at the closed end, but instead this phase change occurs at the open end. Secondly, it is necessary to introduce an end correction at the open end of the pipe. It is found experimentally that the 1D sound waves are reflected back into the pipe by the 3D surrounding atmosphere at an effective distance $\Delta L \sim 0.3D$ beyond the pipe's end, where D is the pipe's diameter. The effective length L'_E of the pipe with one or both ends open is thus slightly longer than L. Standing wave conditions for sound waves in pipes are readily obtained using the approach applied to waves on strings.

The standing wave condition for harmonic sound waves in a pipe closed at both ends is identical to that for a string fixed at both ends as given in Equation (9.10), with $L = n\,\lambda_n/2$ and $n = 1, 2, 3, \ldots$. The same standing wave condition applies to a pipe open at both ends provided L is replaced by L'_E. Finally, for a pipe open at one end and closed at the other, the standing wave condition is $L'_E = n\,\lambda_n/4$ with $n = 1, 3, 5, \ldots$.

Precisely the same stranding wave conditions are obtained using the fluctuating pressure form of the sound wave function given in Equation (9.13). In this description, nodes, at atmospheric pressure, occur at open ends of pipes and antinodes at closed ends. In addition, $180°$ phase shifts occur at antinodes but not at nodes.

Exercise 9.9: Determine the frequency of the third harmonic for standing waves in a pipe of length 1.2 m which is open at one end and closed at the other. What would the frequency be if the tube were open at both ends? Assume that the pipe's diameter is much less than its length, so that the length correction at the open ends can be omitted. Take the speed of sound as 340 m/s.

The standing wave condition for a pipe which has one end open and the other closed, is $L = n\,\lambda_n/4$ with $n = 1, 3, 5, \ldots$. For the third harmonic $n = 3$, and this gives the wavelength of the corresponding standing wave as $\lambda_3 = 4L/3 = 1.6$ m. The frequency is obtained using the relation $f = v/\lambda$, which gives $f = 212.5$ Hz.

If the pipe were open at both ends, the standing wave condition would become $L = n\,\lambda_n/2$ with $n = 1, 2, 3, \ldots$. The wavelength of the third harmonic, with $n = 3$, becomes $\lambda_3 = 2L/3 = 0.8$ m, with frequency $f = 425$ Hz.

Note that that the wavelength halves and the frequency doubles when the open end of the pipe is closed, converting this boundary from an antinode to a node in the displacement of mass elements description.

9.7.4 MUSICAL INSTRUMENTS

Musical instruments, based on standing wave phenomena, are classified as string or wind or percussion devices, and operate over a wide range of frequencies. The size of the instrument determines the frequency range of the musical notes that are produced. The sound wave frequencies are, naturally, linked to the frequencies that can be heard by the human ear. For a young person, this range extends from ~20 Hz to ~20,000 Hz. In middle age, the upper frequency limit starts to decrease, and the highest frequency notes can no longer be heard. Listening to very loud sounds for lengthy periods can accelerate this loss of hearing. The larger an instrument is, the lower the range of frequencies that it can cover. Large pipe organs can produce notes at frequencies as low as 10 Hz, while the lowest note for a flute is near 220 Hz. The analysis of the sounds that can be produced by musical instruments is complicated, particularly for instruments with conical shapes such as saxophones and tubas. Sound spectrum analysers provide detailed information on the sound waves that are produced by the various instruments. Most instruments produce several harmonics simultaneously rather than just one note. Further details are given in specialist articles on this subject.

10 Waves in Higher Dimensions

10.1 INTRODUCTION

As mentioned in Chapter 9, there are various types of waves in nature, including sound waves, water waves, seismic waves, electromagnetic waves, and the recently detected gravitational waves, which are produced during the acceleration of astronomically large masses. An important distinction is made between waves which require a medium in which to propagate, such as sound waves and water waves, and waves which do not need a medium and propagate in free space, as exemplified by electromagnetic waves and gravitational waves. The wavelength ranges of the different types of waves vary considerably. In spite of this important difference, waves in general share common features, with many of the properties described using the same basic formalism. This chapter is largely concerned with 3D sound waves in fluids, but includes a brief mention of the other types of waves.

10.2 ENERGY AND INTENSITY OF 3D SOUND WAVES

Consider a localized source of 3D sound waves situated in an isotropic medium such as air. When the source is activated, it generates waves with spherical wavefronts which travel away from the source at the speed of sound in the medium. The situation is depicted in Figure.10.1. For large transmission distances L the source is effectively a point source.

Sound waves transport energy through the transmitting medium, as shown in Chapter 9 for 1D harmonic sound waves in pipes. For 3D harmonic sound waves, there is again an outward flow of energy through the medium transmitting the waves. However, in contrast to the 1D sound wave in a pipe case, the energy density of 3D waves decreases as the inverse square of the distance r from the source, as the areas of the expanding spherical wavefronts increase. An expression for the energy per unit volume at a large distance from a source can be obtained by extending the 1D harmonic sound wave model used in Chapter 9. Details are given below.

At large distances from a harmonic sound wave source, the wave function expressed in terms of fluctuations in position of a small disk-shaped volume element ΔV is similar to that for a sound wave in a pipe as introduced in Chapter 9. Figure 10.1 illustrates how a portion of a spherically symmetric wave that has passed through a

DOI: 10.1201/9781003485537-10

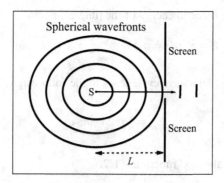

FIGURE 10.1   2D representation of spherical wavefronts propagating outwards from a point source S. If a screen containing a small circular aperture of width d is placed in the path of the wavefronts at a distance L from the source, as shown, then a beam will propagate beyond the screen, and for large L and $d \gg \lambda$ will approximate plane waves over a short distance. A representative disk-shaped volume element in this small quasi-1D region is used in determining the energy per unit volume and the rate at which energy flows through the volume element.

circular aperture in a screen at a large distance L from the source S approximates a 1D sound wave in a limited volume region beyond the screen. Note the similarity of the cylindrical portion of the wave that has passed through the aperture in Figure 10.1 and the wave segment shown in Figure 9.15.

Adapting the 1D form given in Equation (9.14), by replacing x with the wave induced displacement q, which is directed along the radial distance r from the source to the volume element ΔV, gives the required wave function at large r as

$$\varphi(q,t) = q_m \sin(kq - \omega t) \tag{10.1}$$

(The phase angle $\pi/2$ is omitted here, since the phase is arbitrary in the present discussion, which focuses on the time dependence of the displacement of the fluid volume element ΔV.) The kinetic energy of the element is

$$K = \frac{1}{2}m\,v^2 = \frac{1}{2}\rho\,\Delta V\left(\frac{d\varphi}{dt}\right)^2 = \frac{1}{2}\rho\,\Delta V\,\omega^2\,q_m^2\cos^2(kx - \omega t) \tag{10.2}$$

where ρ is the density of the fluid. Following the procedure used for 1D harmonic waves in Chapter 9, it is assumed that the potential energy U is equal to the kinetic energy K, giving

$$U = \frac{1}{2}\rho\,\Delta V\,\omega^2\,q_m^2\cos^2(kx - \omega t) \tag{10.3}$$

The total sound wave induced energy of the fluid element is

$$E = K + U = \rho \, \Delta V \, \omega^2 \, q_m^2 \cos^2 \left(k \, x - \omega t \right) \tag{10.4}$$

Averaging over a complete wave cycle leads to the following result,

$$E_{av} = \frac{1}{2} \rho \, \Delta V \, \omega^2 \, q_m^2 \tag{10.5}$$

(The time average of the $\cos^2$ function is $1/2$.)

The energy density of the propagating sound waves $\varepsilon = E_{av} / \Delta V = \frac{1}{2} \rho \omega^2 \, q_m^2$ thus depends firstly on the mass density ρ of the transmitting medium, secondly on ω^2 the angular frequency squared, and thirdly on the square of the maximum displacement q_m^2, which in turn depends on the distance r from the sound source. For spherical wavefronts, the total volume involved, which corresponds to the sum of all the volume elements in a shell of radius r and thickness Δr, is $4\pi r^2 \, \Delta r$. Since the total energy associated with a propagating spherical wavefront is constant, assuming energy loss mechanisms are negligible, it follows that the energy density, and hence q_m^2, falls off as $1/r^2$.

It is useful to introduce the sound wave intensity I, which is the average transmitted power per unit area perpendicular to the direction of wave propagation at some point a distance r from the sound source. The average power P_{av} is the rate at which sound energy leaves the source averaged over time. For spherical wavefronts $I = \dfrac{P_{av}}{4\pi r^2}$ which shows that the intensity falls off with distance as $1/r^2$. In Equation (10.5), the volume of the element considered can be taken as $\Delta V = A \, v_s \, \Delta t$, where A is the cross-sectional area, v_s is the speed of sound waves in the medium, and Δt is the time taken for a wave to travel the distance Δr, which is the width of a volume element.

Replacing ΔV by $A \, v_s \, \Delta t$ in Equation (10.5) gives $E_{av} = \frac{1}{2} \rho A \, v_s \, \Delta t \, \omega^2 \, q_m^2$, and hence the average power is

$$P_{av} = \frac{1}{2} \rho A \, v_s \, \omega^2 \, q_m^2 \tag{10.6}$$

The intensity then follows as

$$I = \frac{1}{2} \rho \, v_s \, \omega^2 \, q_m^2 \tag{10.7}$$

The SI units of intensity are J s^{-1} m^{-2}, or W m^{-2}.

Sound intensity varies over an exceptionally large range and depends on both the source power and the distance from the source at which the sound waves are detected. The human ear can detect sound waves with an intensity in the range from 10^{-12} W/m^2

to above 1 W/m². It is therefore convenient to use a logarithmic intensity scale, and the following expression for comparing intensities has been introduced:

$$\beta = 10 \times \log_{10} \frac{I}{I_0} \qquad (10.8)$$

Values of β are quoted in decibels (dB). The reference intensity I_0 is taken as 10^{-12} W/m², to coincide roughly with the human detection limit. An intensity $I = 10^{-12}$ W/m² corresponds to $\beta = 0$ dB, while $I = 1$ W/m² gives $\beta = 120$ dB. Using the relation $\Delta p_m = v_s \rho \omega q_m$ from Section 9.7, together with Equation (10.7), leads to $I = \dfrac{(\Delta p_m)^2}{2\rho v_s}$ which shows that the intensity of sound is proportional to the square of the amplitude of pressure fluctuations. The minimum amplitude of pressure fluctuations that can be detected by the human ear is $\Delta p_0 = 2 \times 10^{-5}$ Pa, which is ten orders of magnitude smaller than standard atmospheric pressure. In terms of Δp_m and Δp_0 values, Equation (10.8) is rewritten as

$$\beta = 10 \times \log_{10} \frac{\Delta p_m^2}{\Delta p_0^2} = 20 \times \log_{10} \frac{\Delta p_m}{\Delta p_0} \qquad (10.9)$$

Exercise 10.1: At loud music concerts, the sound level can exceed 100 dB near the stage. What is the intensity of sound corresponding to 100 dB? Calculate the pressure fluctuations amplitude produced in air by this sound. What is the displacement amplitude at 500 Hz? The density of air is $\rho_a = 1.21$ kg/m³ and the speed of sound in air is $v_a = 340$ m/s.

Taking $\beta = 100$ in Equation (10.8) gives $I = 10^{-12} \times 10^{10} = 10^{-2}$ W/m². The amplitude of pressure fluctuations for 100 dB sound is obtained from Equation (10.9) which gives $\Delta p_m = 10^5 \times (2 \times 10^{-5}) = 2$ Pa. The displacement amplitude is given by $q_m = \dfrac{\Delta p_m}{v_a \rho_a \omega} = \dfrac{2}{340 \times 1.21 \times (2\pi \times 500)} = 1.55 \times 10^{-6}$ m. This means that the amplitude of the displacement of air elements for 500 Hz sound waves with $\beta = 100$ dB is approximately one micron.

10.3   INTERFERENCE EFFECTS

10.3.1   INTERFERENCE IN 1D

Wave interference effects are observed when two or more waves from different sound sources, with fixed phase relationships, are superposed. A fixed phase relationship between two or more wave sources with the same frequency is expressed compactly by stating that the outgoing waves are coherent. The special case of two harmonic waves with the same amplitude and frequency travelling parallel to the x-axis will be considered first, but the approach is readily extended to other similar situations. The basic ideas are much the same as those used in the description of standing waves in pipes, involving a single source of harmonic waves with reflections of the waves at the two ends of a pipe as described in Section 9.7. Consider two waves which have equal amplitudes and frequencies with wave functions given in terms of the pressure variation by $\Delta p_1 = \Delta p_m \sin(k\,x - \omega t + \theta)$ and $\Delta p_2 = \Delta p_m \sin(k\,x - \omega t)$. The phase difference θ is constant. If the waves interfere, the superposition principle gives the resultant wave function as

$$\Delta p = \Delta p_1 + \Delta p_2 = \Delta p_m \left[\sin(k\,x - \omega t + \theta) + \sin(k\,x - \omega t) \right] \qquad (10.10)$$

Making use of the trigonometric identity $\sin \alpha + \sin \beta = 2\cos \dfrac{\alpha - \beta}{2} \sin \dfrac{\alpha + \beta}{2}$ gives

$$\Delta p = 2\Delta p_m \cos\left(\frac{\theta}{2}\right)\sin\left(k\,x - \omega t + \frac{\theta}{2}\right) \qquad (10.11)$$

The amplitude of the resultant wave depends on the phase difference θ and is given by $2\Delta p_m \cos(\theta/2)$. Constructive interference occurs for $\cos(\theta/2) = 1$ corresponding to $\theta = 0, 2\pi, 4\pi, 6\pi, \ldots$ (zero and even multiples of π), while destructive interference occurs for $\cos(\theta/2) = 0$ when $\theta = \pi, 3\pi, 5\pi, 7\pi, \ldots$ (odd multiples of π).

If the sources of the interfering waves are initially in phase, with $\theta = 0$, but the waves travel different distances x_1 and x_2, with $\Delta x = x_2 - x_1$, then the phase difference becomes $\theta = k\,\Delta x = 2\pi\,\Delta x/\lambda$, which is just like the travelling waves on a string that are discussed in Chapter 9. For example, when $\Delta x = \lambda$ it follows that $\theta = 2\pi$.

10.3.2 Interference in 3D

The interference conditions given above for waves travelling parallel to the x-axis are readily extended to 3D waves. Consider two coherent sources 1 and 2 separated by a distance d as represented in 2D cross-section in Figure 10.2. An absorbing barrier is placed behind the sources. Hemispherical waves of wavelength λ propagate as shown.

As shown in Figure 10.2, constructive interference occurs when wave crests from the two sources coincide. The interference pattern has a central maximum, with secondary maxima radiating outwards above and below the central maximum. The condition for interference maxima to occur is given in terms of the path difference Δr as $\Delta r = n\,\lambda$ with $n = 0, 1, 2, \ldots$. With increasing radii of the wavefronts, the maxima become aligned along the directions indicated by the dash lines.

At sufficiently large distances from two harmonic wave sources, the condition for interference maxima to occur has a simple compact form. Let interference of waves from two wave sources, S_1 and S_2 separated by d, occur at a point P which is at a large distance $r \gg d$ from the sources. The angle θ specifies the orientation with respect to

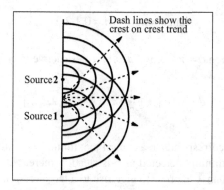

FIGURE 10.2 Two harmonic wave sources separated by a distance d generate coherent 3D waves of wavelength λ. The two sets of travelling waves are depicted in 2D at a fixed time t. Constructive interference occurs in directions along which wave crests from the two sources reinforce. The dash lines show the trend in interference maxima locations with increasing radii of wave crests. Destructive interference occurs in directions for which crests and troughs overlap (not shown). The interference pattern consists of a central maximum flanked by secondary maxima. The path difference Δr for the identified maxima to occur is given by $\Delta r = n\lambda$ with $n = 0$, 1, and 2 .

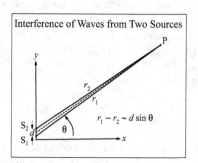

FIGURE 10.3 Spherical waves of wavelength λ are generated by two coherent harmonic sources S_1 and S_2, which are separated by a distance d. Interference effects are detected at a point P situated at a distance r_1 from S_1 and r_2 from S_2. If d and λ are much less than both r_1 and r_2, then trigonometry gives $r_2 - r_1 \approx d\sin\theta$. In this limit, the condition for constructive interference becomes, to a good approximation, $d\sin\theta = n\lambda$ with $n = 0,1,2,3,\ldots$.

the x-axis of a line drawn from the midpoint between S_1 and S_2 to point P as shown in Figure 10.3.

At sufficiently large distances from the two sources, compared to their separation d, the radius vectors $\mathbf{r_1}$ and $\mathbf{r_2}$ of the two wavefronts at P are travelling close to parallel. As can be seen in Figure 10.3, waves from S_1 travel a distance $\Delta r = d\sin\theta$ further than do waves from S_2 in reaching P. The phase difference between the waves is therefore $\phi = 2\pi\,\Delta r/\lambda = 2\pi\,d\sin\theta/\lambda$. The set of phase values $\phi = 0, 2\pi, 4\pi, 6\pi,\ldots$ required for constructive interference leads to the relationship:

$$d \sin \theta = n \lambda \quad (n = 0, 1, 2, 3, \ldots) \tag{10.12}$$

Similarly, the condition $\phi = \pi, 3\pi, 5\pi, 7\pi, \ldots$ for destructive interference gives

$$d \sin \theta = \left(n + \frac{1}{2} \right) \lambda \quad (n = 0, 1, 2, 3, \ldots) \tag{10.13}$$

The condition $\theta = 0$ corresponds to a central maximum in sound intensity, with alternating maxima and minima detected as θ is steadily increased. Note that d must be significantly larger than λ in order to give rise to a large set of n values.

Exercise 10.2: Two loudspeakers, mounted 1.5 m above ground level and separated by a horizontal distance of 3.0 m, send out coherent harmonic sound waves at a fixed frequency. A sensor, which is used to detect the resultant sound wave amplitude as a function of position in a horizontal plane, moves along an arc of radius 30 m centred on the loudspeaker pair midpoint. If the third maximum from the centre of the interference pattern is found to be at an angle of 30° away from the central maximum direction, what is the wavelength of the sound?

Using Equation (10.12) with $n = 3$ gives $\lambda = (d \sin \theta)/3 = 3 \times 0.5/3 = 0.5$ m . Taking the speed of sound in air as $v_a = 340$ m/s gives the frequency as $f = v_a/\lambda = 680$ Hz.

10.4   BEATS

Consider the interference of two sound waves of equal amplitude, but slightly different angular frequencies ω_1 and ω_2, with $\omega_2 > \omega_1$. Instead of observing time-independent interference effects, as discussed in Section 10.3, the resultant amplitude exhibits time-dependent oscillations. Adapting Equation (10.10) gives the amplitude of the interfering waves as $\Delta p_R = \Delta p_1 + \Delta p_2 = \Delta p_m \left[\sin\left(k_1 \, x - \omega_1 \, t + \theta\right) + \sin\left(k_2 \, x - \omega_2 \, t\right) \right]$. In order to simplify the equations, it is convenient to choose $x = 0$ as the position at which the resultant wave is examined by a listener. The two waves are taken to be in phase at $t = 0$ with $\theta = 0$. Making use of the trigonometric identity $\sin \alpha + \sin \beta = 2 \cos \dfrac{\alpha - \beta}{2} \sin \dfrac{\alpha + \beta}{2}$ in order to carry out the summation, leads to

$$\Delta p_R = 2 \Delta p_m \cos\left(\frac{1}{2}\left(\omega_2 - \omega_1\right)t \right) \sin\left(\frac{1}{2}\left(\omega_1 + \omega_2\right)t \right) \tag{10.14}$$

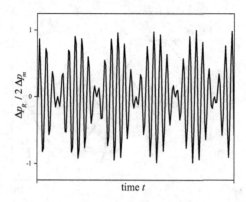

FIGURE 10.4 A representative beat pattern is produced by the superposition of two harmonic waves of equal amplitude, but with slightly different frequencies, as given by Equation (10.15).

Defining the frequency difference as $\Delta\omega = \left(\omega_2 - \omega_1\right)$ and the average frequency as $\omega_{av} = \dfrac{1}{2}\left(\omega_1 + \omega_2\right)$, and substituting in Equation (10.14) gives

$$\Delta p_R = 2\Delta p_m \cos\left(\frac{\Delta\omega\, t}{2}\right)\sin\left(\omega_{av}\, t\right) \qquad (10.15)$$

A plot depicting the variation of $\Delta p_R / \Delta p_m$ with time is shown in Figure 10.4. As can be seen from Equation (10.15), the resultant is an amplitude modulated sine wave of angular frequency ω_{av} and time-dependent amplitude $2\Delta p_m \cos(\Delta\omega\, t/2)$. The interfering waves alternate between constructive and destructive interference with the passage of time.

A listener hears an alternating sound intensity, which is proportional to the square of the amplitude of the beat frequency. The beat frequency heard by a listener is given by $f_B = 2\pi\,\Delta\omega$, which is twice the amplitude modulation frequency because the intensity increases to a maximum twice per modulation cycle. Musical instruments can be tuned using the beat phenomenon as the basis for comparing a note on an instrument with that of a frequency standard such as a tuning fork.

10.5 FOURIER ANALYSIS

An interesting application of the superposition of waves of different frequencies is known as Fourier analysis. The procedure involves adding together harmonic waves of selected frequencies and amplitudes in order to generate a periodic wave of a particular form, such as a triangular wave or a square wave. The harmonic wave components are known as Fourier components. As an illustrative example consider the case of a square wave of period T with the form shown in Figure 10.5.

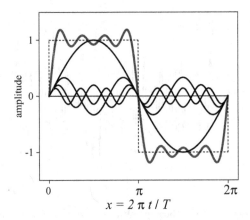

$$x = 2\pi t / T$$

FIGURE 10.5 The dash line in the figure shows the amplitude variation over one cycle for a square wave of period T. The light curves are the first four Fourier components in the Fourier series for this waveform. The heavy line, which is the sum of the Fourier components, approximates the required shape. The inclusion of additional higher order components would improve the agreement.

Equation (10.16) gives the first four components of the Fourier series for a square wave as a function of $x = 2\pi t/T$:

$$f(x) = \frac{4}{\pi}\left(\frac{\sin x}{1} + \frac{\sin 3x}{3} + \frac{\sin 5x}{5} + \frac{\sin 7x}{7} + \ldots\right) \qquad (10.16)$$

The sum of the components is also shown in Figure 10.5. A large number of terms with steadily increasing frequencies are required in order to obtain a close approximation to the square wave. Fourier analysis provides an instructive way of examining all periodic waveforms in terms of their constituent harmonics.

10.6 DOPPLER SHIFTS

Doppler frequency shifts of mechanical harmonic waves travelling through a medium are observed when the wave source and/or the wave detector are in motion with respect to the medium. This phenomenon is known as the Doppler effect. The present discussion focuses on sound waves that travel at a fixed speed in the transmitting medium, such as air near the Earth's surface. Doppler shifts also occur with electromagnetic waves and, for example, are important in interpreting spectroscopic measurements in astrophysics. Earthbound applications include radar speed guns for measuring the speed of motor vehicles on highways. It is important to note that the Doppler effect expressions for sound waves are similar to those for electromagnetic waves, but are not the same. In contrast to sound waves, electromagnetic waves do not require a medium for their transmission and travel at the speed of light in vacuum as manifested in the special theory of relativity.

10.6.1 Moving Source

Consider a source of 3D harmonic sound waves, with frequency f, which is initially at rest in a transmitting medium such as air. Taking the speed of sound in the medium as v, the corresponding wavelength is $\lambda = v/f$. If the source S is moved at speed v_S towards the detector, labelled D_1 in Figure 10.6, then the frequency f' measured at the detector is greater than f. This Doppler frequency shift is due to the bunching up of wavefronts ahead of the moving source as shown in Figure 10.6.

Bunching up of wave crests leads to an *effective wavelength* $\lambda' = (v - v_S)T$ where $T = 1/f$ is the oscillation period at the source. It follows that $(v - v_S)T$ is the peak-to-peak separation of wave crests moving away from S. Also, the effective wavelength is given by $\lambda' = v/f'$, because the actual wave speed v in the medium does not change. Using the two expressions for λ' gives $(v - v_S)/f = v/f'$, and rearranging leads to the following expression for the Doppler shifted frequency at D_1

$$f' = \frac{v}{v - v_S} f \qquad (10.17)$$

For the detector labelled D_2, from which the source S is moving away, the minus sign in the denominator of Equation (10.17) changes to a plus sign.

10.6.2 Moving Detector

If a detector D_1 is moved towards a stationary sound source S at speed v_D, then there is no change in the sound wavelength λ because there is no wave bunching effect. However, the moving detector encounters an increased number of wave crests in

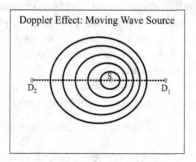

FIGURE 10.6 A moving source S generates spherical harmonic sound waves as it travels towards a fixed detector D_1 along the path indicated by the dotted line, as shown in this 2D representation of successive wave crests. The wave crests ahead of the moving source are bunched up, leading to an increase in the frequency detected at D_1. In contrast, the detector D_2 behind the moving source detects a decrease in frequency.

a given time interval than it would have if stationary, and this results in a shift in detected frequency from f to f'. The wavelength is written as $\lambda = \left(v + v_D\right)/f'$ where $\left(v + v_D\right)$ is the *effective speed* of the waves in the detector's frame of reference. In the source's frame of reference $\lambda = v/f$. Equating the two expressions for λ, and rearranging, gives the shifted frequency as

$$f' = \frac{v + v_D}{v} f \tag{10.18}$$

If D moves away from the source, then the plus sign in the numerator of Equation (10.18) becomes a minus sign.

When both S and D are in motion, it is necessary to combine Equation (10.17) with Equation (10.18) to obtain the general Doppler shift expression:

$$f' = \frac{v \pm v_D}{v \mp v_S} f \tag{10.19}$$

The choice of signs in the numerator and denominator is determined by the direction of motion of the source and detector with respect to one another. Movement towards each other leads to an increase in frequency, while movement apart leads to a decrease.

Equations (10.17) and (10.18) apply to situations in which the speed of the source or the detector is much lower than the speed of sound in the transmitting medium. As discussed below, interesting effects arise when an object, like a jet plane, travels at supersonic speeds.

Exercise 10.3: A police car travelling at 120 km/h is equipped with a siren which generates sound that alternates between two frequencies, 635 and 912 Hz in the police car's frame of reference. What range of frequencies will be heard by a stationary pedestrian at the side of the road who watches the approach of the police car? How will the frequency range change as the police car moves down the road beyond the observer?

At some instant, the police siren emits sound waves of frequency f as it moves at speed v_S towards the observer who hears the siren sound at frequency $f' = \dfrac{v}{v - v_S} f$ as given by Equation (10.17). Taking the speed of sound in air as $v = 340$ m/s, and converting the car's speed units from km/h to m/s, gives for the low end of the frequency range $f'_L = \dfrac{340}{340 - 33.3} \times 635 = 704$ Hz. The upper end of the Doppler shifted frequency range is obtained in a similar fashion as $f'_U = \dfrac{340}{340 - 33.3} \times 912 = 1011\,\text{Hz}$. As the police car moves away from the

observer, the frequencies heard are given by $f' = \dfrac{v}{v + v_S} f$, with the minus

sign in the denominator changed to a plus sign. Substituting numbers gives $f'_L = 578\,\text{Hz}$ and $f'_U = 831\,\text{Hz}$.

10.6.3 SHOCK WAVES

If the speed of a wave source is increased until it exceeds the speed of sound in the transmitting medium, then sound waves cannot propagate ahead of the source. A shock wave develops behind the source, as depicted in Figure 10.7, with a cone-shaped boundary known as the Mach cone forming behind the fast-moving wave source.

The cone shape of the shock wave is determined by the ratio of the distances travelled by the supersonic source and that covered by sound waves as a function of time. If, in Figure 10.7, O is taken as the starting point for sound waves travelling from O to Q in a certain time interval t, then, in the same time interval, the source travels a distance $OP = v_S t$ which is greater than the distance $OQ = vt$. As a result, a shock wave is generated by the overlapping of wave crests along the Mach cone boundary. The sonic boom produced by supersonic aircraft is an example of this effect. The half-angle of the Mach cone is thus given by $\sin\theta = v/v_S$.

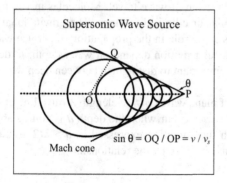

FIGURE 10.7 A wave source, travelling at a speed v_S, which exceeds the speed of sound v in air, generates a trailing shock wave. The loci of the overlapping wave crests form a cone, which is known as a Mach cone. An observer in the path of the Mach cone experiences an upward surge in pressure as the shock wave passes, followed by a drop in pressure, before the pressure levels return to normal. The half angle θ of the Mach cone is given by $\sin\theta = v/v_S$.

Exercise 10.4: Determine the predicted Doppler effect behaviour as a wave source increases its speed towards the speed of sound in the transmitting medium.

From Equation (10.17), the ratio of the Doppler shifted frequency to the source frequency is given by $f'/f = v/(v - v_s)$. As $v_s \to v$, the ratio f'/f becomes exceptionally large, tending to infinity. Detection of Doppler shifts is no longer possible in this limit. For $v_s = v$, the waves from the source travel at the same speed as the source leading to a buildup of pressure as the wavelets overlap each other.

10.7 WATER WAVES

Familiar characteristic behaviours of water waves include the spread of ripples on the surface of a pond into which a pebble has been thrown, and the breaking of wind-generated ocean waves on the shore. Harmonic water waves can be generated in a controlled way using devices called ripple tanks, in which vibrating sources dip into the water surface. By selecting the wave source geometry to be either extended or point like, it is possible to generate either plane waves or circular waves. In addition, a single source or multiple sources can be used, allowing a variety of overlapping wave patterns to be formed. As an example, two point-like sources produce an interference pattern similar to the pattern shown in Figure 10.2. The speed at which water waves propagate depends on the depth of water in the container across which the waves travel. This feature is of particular importance in understanding wave behaviour on lake and ocean surfaces.

In contrast to the longitudinal nature of sound waves in air, the surface waves in water involve both longitudinal and transverse displacements of water molecules with respect to the direction of wave travel. Molecules in volume elements near the surface execute circular or elliptical motions. The gravitational force acting on the water elements plays a key role in the propagation of surface waves. Both the speed and the form of molecular motion depend on water depth. In dealing with this type of wave motion, it is important to distinguish between deep water waves and shallow water waves.

Consider a train of plane waves of wavelength λ travelling at speed v over a water surface. Deep water waves occur when the depth d is comparable to, or larger than, the wavelength, with the condition expressed as $d > \lambda/2$. Taking $g = 9.8$ m/s^2, the wave speed in this case is given by the relationship

$$v = \sqrt{\frac{g\,\lambda}{2\pi}} \approx 1.25\sqrt{\lambda} \tag{10.20}$$

In contrast, shallow waves require that the depth be much less than the wavelength, and the necessary condition is taken as $d < \lambda/20$. In this limit, the speed expression is

$$v = \sqrt{g\,d} = 3.13\sqrt{d} \tag{10.21}$$

While the derivations of Equations (10.20) and (10.21) are not given here, it is readily seen that the expressions are dimensionally correct.

The values of λ for wind-driven, deep water waves lie in the range 50–150 m, corresponding to wave speeds of 9–15 m/s. As deep water waves approach the shore, they transition into shallow water waves. Consequently, as the speed decreases, the wavelength decreases, while the wave amplitude increases and the shape changes. The kinetic energy transported by a wave approaching the shore does not decrease significantly, and this causes the buildup in amplitude as the speed drops. Because of the variation in d between the bottom and top of the wave, the top travels more rapidly than the bottom leading to the well-known breaking behaviour of sea waves.

Seismic waves, known as *tsunamis*, are particularly dangerous and can cause immense damage when they strike coastal regions. Earthquakes deep below the ocean floor cause these waves, which have exceptionally long wavelengths $\lambda > 100$ km, with long periods $T \sim 20$ min. Tsunami waves are always a shallow water wave because of their large wavelengths.

Exercise 10.5: Estimate the speed of a tsunami in the Pacific Ocean, which has an average depth of 4000 m.

Equation (10.21) for shallow water waves gives $v = 3.13\sqrt{4000} = 198$ m/s. This speed, $v \sim 710$ km/h, is comparable to that of a passenger jet aircraft.

10.8 ELECTROMAGNETIC WAVES

Electromagnetic waves were predicted in 1865 by James Clerk Maxwell. Maxwell based his prediction on his set of equations, which describe electromagnetic phenomena. Roughly two decades later, Heinrich Hertz provided experimental verification of Maxwell's prediction. Rapid progress in determining the properties of these waves followed. The present brief introduction to classical electromagnetic waves simply points out common features that electromagnetic waves share with other types of waves, together with important differences in behaviour.

Electromagnetic waves do not require a medium in which to propagate and travel at the speed of light $c = 2.99792458 \times 10^8$ m/s in vacuum. The electromagnetic spectrum spans a wide range of wavelengths from gamma rays (10^{-15} m) and X-rays, through visible light, to microwaves and very long wavelength radio waves (10^8 m). Wave frequencies f, which are related to the wavelengths λ by the equation $c = f \lambda$, correspondingly have values ranging from 10^{24} Hz to Hz. A variety of electronic devices have been developed that make use of this wide spectral range.

Based on Maxwell's contributions and the work of others, a plane-polarized harmonic electromagnetic wave propagating parallel to the x-axis in a Cartesian frame involves time-varying electric field, E, and magnetic field, B, components aligned orthogonally to x. If E is directed along y, then B is directed along z. The harmonic wave functions are written as $E_y = E_m \sin(k x - \omega t)$ and $B_y = B_m \sin(k x - \omega t)$, with $k = 2\pi/\lambda$ and $\omega = 2\pi/T = 2\pi f$. In addition, the two amplitudes E_m and B_m are closely related, with $E_m/B_m = c$ and the two components in phase. The harmonic wave functions for each component are similar to the wave function of a harmonic wave on a string as discussed in Chapter 9. Note that the chosen forms correspond to polarized waves, with the electric field confined to the xy plane and the magnetic field in the xz plane. Interestingly, it is the electric component which is responsible for many of the observable physical effects that are detected using electromagnetic radiation.

Using the forms given above for the electric and magnetic field wave functions, it follows that the phenomena which are found with mechanical waves on strings, and with sound waves in air, have their counterparts in electromagnetic wave phenomena. For example, the superposition of two light waves of the same frequency give rise to interference effects. A major difference between mechanical waves and electromagnetic waves is the difference in the observed transmission speeds. Electromagnetic waves travel at the speed of light, which is invariant for observers in different inertial reference frames that are in relative motion. In contrast, sound waves are seen to travel at different speeds for observers in relative motion with respect to the transmitting medium.

While the classical wave description is successful for describing many of the observed properties of electromagnetic radiation, it is necessary to extend the description in order to account for quantum transitions involving the interaction of radiation with matter at the atomic level. This major alteration to the wave picture is made by introducing the photon. Photons are fundamental particles that travel at the speed of light and carry momentum and energy. A collection of these particles constitutes the wave of classical physics. In the quantum physics description, the wave function is linked to the probability density of photons, which make up the wave, expressed in terms of space and time coordinates. Further details are given in books on quantum mechanics.

11 Basics of Thermal Physics

11.1 INTRODUCTION

Thermal physics began to be established as an important subject in the nineteenth century, in large part due to the technological advances that accompanied the industrial revolution. In particular, it was the design and construction of heat engines that provided a major stimulus for the investigation of thermal processes. The concepts of work and energy from classical mechanics feature in a natural way in the description of thermal phenomena. A related concept, that of heat energy, became essential during the development of the subject. This chapter provides an introduction to the subject called thermodynamics.

Thermodynamics provides a description of the processes that occur in macroscopic systems consisting of very large numbers of particles. No attempt is made to connect the macroscopic behaviour that is observed to the unseen microscopic constituents. A related subject, called kinetic theory, does provide a semi-classical connection to the microscopic description based on classical models. Quantum mechanics is used in developing statistical physics, which provides a deeper connection to thermodynamics.

In introducing the subject, it is convenient to consider what is called the ideal gas as a simple but important system of interest. The equation of state for an ideal gas connects pressure, volume, and temperature in a compact way. Other systems are introduced, where appropriate, to broaden the discussion. While the pressure and volume of fluids are familiar concepts from the discussion given in Chapter 7, the scientific role of temperature is less familiar and is therefore dealt with in detail in Section 11.2. In addition to the Celsius and Fahrenheit temperature scales, which are well known, the absolute or Kelvin scale is defined.

The internal energy of a system together with the work done on or by the system are important in formulating the laws of thermodynamics. As mentioned above, it is also necessary to introduce heat energy, which can be transferred in thermal interaction processes. Heat flow across the boundary of a system involves the transfer of energy at the microscopic scale. While thermodynamics does not consider the microscopic nature of heat transfer processes, it is instructive to consider basic kinetic theory in order to gain insight into these processes. An overview of the kinetic theory of gases is given in Section 11.3.

DOI: 10.1201/9781003485537-11

11.2 TEMPERATURE AND THE IDEAL GAS LAW

Thermometers that make use of the thermal expansion properties of liquids typically consist of a reservoir of liquid attached to a capillary tube. The length of the liquid column gives a measure of the temperature of the liquid in the reservoir. Devices of this type are generally calibrated using Celsius or Fahrenheit temperature scales, which are familiar from everyday life. Both scales fix two calibration points, one for melting ice called the ice point, and the other for the boiling point of water in contact with steam at a pressure of one atmosphere called the steam point. In the Celsius scale, the ice point is fixed as 0°C and the steam point at 100°C with linearly marked degree subdivisions. The Fahrenheit scale uses 32°F for the ice point and 212°F for the steam point. The liquids used in thermometers are mercury for common purposes and alcohol for temperatures below the ice point.

Liquid in glass thermometers use the thermal expansion of a liquid in establishing temperature scales. Many other types of thermometers exist including gas thermometers, electrical resistance thermometers, and magnetic thermometers. These devices are all based on changes in the physical properties with temperature of the system used and require calibration. For scientific purposes, it is convenient to use the Kelvin scale, also known as the absolute temperature scale, which is introduced below. In introducing the Kelvin scale, it is instructive to consider the physical properties of gases as a starting point.

The pressure of a gas in a container depends on the container volume, the quantity of gas in the container, as measured in moles, and finally the temperature. Experiments carried out by many workers, beginning in the seventeenth century, established these dependences. These investigations led to concepts and laws associated with the names of Boyle, Charles, Gay-Lussac, and Avogadro. It became clear that the basic laws could be unified in what is called the ideal gas law, which has the following form:

$$P\,V = n\,R\,T \tag{11.1}$$

In Equation (11.1), P is the pressure, V the volume, n the number of moles of gas, T the absolute temperature, and R a constant called the gas constant with approximate value $R = 8.314\,\mathrm{J\,mol^{-1}\,K^{-1}}$. The term ideal gas refers to a gas in which there are negligibly small interactions between the constituent particles (atoms or molecules) making up the gas. Examples of gases which are good approximations to an ideal gas are the noble gases helium and argon. Many other gases, including hydrogen and nitrogen, also approximate ideal gases over a wide range of pressures and temperatures. While P and V are familiar quantities, with units Pa and m^3 respectively in SI units, it is necessary to introduce the quantities n and T.

The number of moles of a gas is given by $n = M/M_A$ where M is the mass of gas in a container and M_A is the molar mass. In terms of Avogadro's number N_A, the mole number is $n = N/N_A$, where N is the number of atoms or molecules, collectively referred to as particles, in the container. Avogadro's number has the approximate value $N_A = 6.022 \times 10^{23}$ particles/mol. The molar mass is $M_A = N_A\,m$, with m the particle

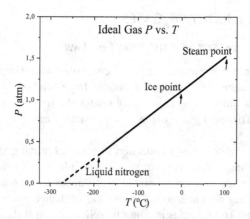

FIGURE 11.1 Representative pressure P (atm) versus temperature (°C) plot for an ideal gas. The volume is $V = 2$ L and $n = 0.1$ mol. Below the liquid nitrogen point, at −190°C, the straight line is extrapolated to zero pressure using the dash line. This occurs at −273.15°C.

mass in atomic mass units (AMU), denoted u, where $1\ u = 1/N_A = 1.660 \times 10^{-24}$ g. The carbon-12 atom is *defined* to have a mass of 12 u. Note that for gases it is conventional to express the molar mass M_A in units of g rather than kg. Avogadro's number is then the number of particles per g-mol.

In order to introduce the absolute temperature T, it is helpful to consider the pressure variation with temperature of a fixed quantity of an ideal gas in a container of fixed volume. Figure 11.1 shows a representative plot of P (atm) versus T (°C).

The pressure of the ideal gas tends to zero at −273°C. This point is a natural choice for absolute zero temperature, $T = 0$ kelvin (K). For the absolute temperature scale, or kelvin scale, degrees are chosen to be equal to those in the Celsius scale. High precision measurements give 0 K as −273.15°C. The ice point, which is fixed as the triple point for solid ice, liquid water, and water vapour to coexist is thus 273.15 K.

The ideal gas law (which is also known as the ideal gas equation of state) as given in Equation (11.1) is extremely useful in thermal physics because it provides a simple and precise description of the behaviour of an ideal gas as a function of the P, V, and T conditions.

Exercise 11.1: Determine the molar volume of an ideal gas for $T = 273.15$ K and $P = 1$ atm. 1 atm $= 1.01325 \times 10^5$ Pa.
From Equation (1.1), $V = n\ R\ T/P = 1 \times 8.314 \times 273.15/101325 = 0.0224$ m³. The molar volume, in litres, is 22.4 L.

11.3   KINETIC THEORY OF GASES

11.3.1   INTERNAL ENERGY AND THE IDEAL GAS LAW

In contrast to thermodynamics, which is not concerned with the microscopic nature of macroscopic systems (i.e. systems that contain large numbers of particles comparable to Avogadro's number), kinetic theory relates the macro and micro properties of such systems. The ideal gas provides a simple and useful example of the kinetic theory approach.

The ideal gas is modelled as a collection of particles moving about inside a container and undergoing collisions with the walls. While the particles are assumed to be noninteracting when they are separated by distances larger than the sum of their effective radii r, they do undergo collisions and exchange momentum when their centres are $2r$ apart. The particles may not be spherical, but if they are not then they rotate rapidly about their centres of mass so that they approximate spheres.

The internal energy E of a physical system is of central importance in the thermodynamic description of processes in which the system is involved. For an ideal gas, the kinetic theory classical model relates the average kinetic energy of a particle ε_k to the absolute temperature. This is done by deriving the ideal gas equation using the ideal gas microscopic model, as shown below. The total energy of the gas is then given by $E = N\,\varepsilon_k$ where N is the total number of particles in the gas. Note that there is no potential energy contribution for an ideal gas because of the vanishingly weak interactions between particles. Variations in gravitational potential energy are also assumed to be unimportant if the gas container is kept at a fixed position in relation to the Earth's surface. Collisions between molecules are ignored although this point is taken up later. Note that the term particle has now been replaced by molecule, which is used, collectively, to describe both atoms such as helium (He) and molecules such as nitrogen (N_2).

Figure 11.2 illustrates in 2D the collision of a gas molecule with the wall of a container.

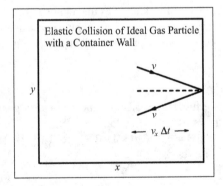

FIGURE 11.2   A molecule in an ideal gas makes an elastic collision with a wall of the gas container. The momentum along x changes by $2m\,v_x$. During an interval of time Δt, many molecules collide with the wall and exert a force on it.

In order to obtain an expression for the pressure of the gas, it is instructive to start by considering a subset of molecules moving with speed v_x in the $+x$ direction. It is assumed that the container wall is smooth and that collisions with it are elastic. The momentum change of a molecule along x is $\Delta p = 2m\, v_x$. Over a time Δt, the fraction of the molecules that strike the wall of area A is given by $f = \frac{1}{2} A\, v_x\, \Delta t/V$. This is because the maximum distance that these molecules will travel is $v_x\,\Delta t$, so all the molecules inside the volume $A\, v_x\,\Delta t$ could strike the surface. The factor $\frac{1}{2}$ is introduced to allow for molecules travelling in either the $+x$ or $-x$ directions. The change in momentum for a molecule colliding with the boundary wall is therefore

$$\Delta p = \left(\frac{A\, v_x\, \Delta t}{2V} \right) \times (2m\, v_x) = A\left(\frac{m\, v_x^2}{V} \right) \Delta t.$$

From Newton's second law, the force F_s produced by a single molecule hitting the boundary wall is obtained as the rate of change of momentum of the molecule considered. This gives $F_s = \dfrac{\Delta p}{\Delta t} = \dfrac{A}{V} m\, v_x^2$. The total force on the wall produced by all the molecules, labelled $i = 1$ to N, is $F = \dfrac{A}{V} m \sum_{i=1}^{N} v_i^2$. Introducing the mean square speed component along x as $v_x^2 = \dfrac{1}{N} \sum_{i=1}^{N} v_i^2$, and substituting in the expression for F gives $F = \dfrac{A}{V} N\, m\, v_x^2$. Since the x, y and z components of the mean square speed v^2 are equal in 3D, as a consequence of symmetry, it follows that $v^2 = v_x^2 + v_y^2 + v_z^2 = 3v_x^2$. The pressure $P = \dfrac{F}{A}$ of the gas in terms of the mean translational kinetic energy $\varepsilon_k = \frac{1}{2} mv^2$ is given by

$$P = \frac{1}{3} N \frac{m\, v^2}{V} = \frac{2}{3} N \frac{\varepsilon_k}{V} \qquad (11.2)$$

Equation (11.2) connects the microscopic (N, ε_k) and macroscopic (P, V) descriptions of an ideal gas.

A comparison of the ideal gas law in Equation (11.1) with the kinetic theory form in Equation (11.2) shows that $n\, RT = \dfrac{2}{3} N \varepsilon_k$, which, using $N = n\, N_A$ and introducing $k_B = R/N_A$, becomes

$$\varepsilon_k = \frac{3}{2} k_B T \qquad (11.3)$$

The new constant k_B is called Boltzmann's constant with approximate value 1.381×10^{-23} J/K. Equation (11.3) relates the mean translational kinetic energy of a single ideal gas molecule to the absolute temperature. Since $\varepsilon_k = \frac{1}{2} m\, v^2 = \frac{3}{2} m\, v_x^2$, the factor of three halves arises because there are three contributions to the kinetic energy of a molecule in 3D space. For a *monatomic* ideal gas of N molecules at temperature T the total energy E is simply the total translational kinetic energy of all the molecules. There is no contribution from the potential energy for an ideal gas. The energy is therefore given by

$$E = N\, \varepsilon_k = \frac{3}{2} N\, k_B\, T = \frac{3}{2} n\, R\, T \tag{11.4}$$

with R the gas constant and n the number of moles. This is an important result in considering the thermodynamic properties of a monatomic ideal gas. Note that for polyatomic molecules there are other contributions to the energy from intramolecular rotational and vibrational motions. Collectively, the various energy contributions are associated with what are called degrees of freedom.

Exercise 11.2: Determine the internal energy of 0.2 moles of helium gas at 20°C.
From Equation (11.4), the energy is given in terms of the gas constant by
$$E = \frac{3}{2} n\, RT = \frac{3}{2} \times 0.2 \times 8.314 \times 293 = 731 \text{ J.}$$

From Equation (11.3), it is possible to obtain an expression for the root mean square speed $v_{rms} = \sqrt{v^2}$ of the molecules in an ideal gas. Equation (11.3) is written as $\varepsilon_k = \frac{1}{2} m\, v^2 = \frac{3}{2} k_B\, T$, and, with m in u, this gives

$$v_{rms} = \sqrt{\frac{3k_B T}{m}} = 158 \times \sqrt{\frac{T}{m}} \text{ m/s} \tag{11.5}$$

Figure 11.3 is a plot of v_{rms} for helium ($m = 4.0$ u) and neon ($m = 20.2$ u) versus T (K).
In an ideal gas at pressures which are not very low, the molecules make frequent collisions with one another and travel small distances between collisions. It is useful to introduce a quantity called the mean free path l, which is a measure of this mean collision distance. Consider a gas of N molecules in a container of volume V. For molecules of effective *diameter a*, two of the molecules undergo collision when their centres are a distance a apart. Collision processes are treated by expanding a selected molecule to twice its size, with *radius a*, and shrinking all the other molecules to geometrical points. The enlarged selected molecule travels at an average speed v

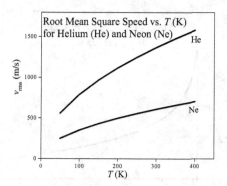

FIGURE 11.3 Root mean square molecular speed versus T (K) for helium and neon ideal gases.

between collisions and sweeps out a volume $V_s = \pi a^2 v \tau$ in a time τ. The number of molecules encountered by the expanded molecule is given by $N_s = \dfrac{V_s}{V} N$. For a single collision $N_s = 1$, and the corresponding collision time is $\tau = \left(\dfrac{1}{\pi a^2 v}\right) \times \left(\dfrac{V}{N}\right)$. This simple approach gives the mean free path as $l = v\tau = \left(\dfrac{1}{\pi a^2}\right) \times \left(\dfrac{V}{N}\right)$. Using the ideal gas law in the form $P\,V = N\,k_B\,T$, and substituting for V/N in the expression for τ, gives $\tau = \left(\dfrac{1}{\pi a^2 v}\right) \times \left(\dfrac{k_B T}{P}\right)$. Introducing a correction factor $1/\sqrt{2}$, which allows for the motion of the other molecules, the mean free path becomes

$$l = v\,\tau = \frac{1}{\sqrt{2}\pi a^2} \frac{k_B\,T}{P} \qquad (11.6)$$

Figure 11.4 shows a semi-logarithmic plot of the mean free path of molecules in helium at 295 K as a function of pressure in the range 0 to 10 atm. The kinetic diameter of He molecules is 0.26 nm. The path length increases dramatically as the pressure drops towards 0 atm.

Note that the mean free path length is inversely proportional to the pressure. In a high vacuum system, at a pressure of 10^{-5} Pa, the mean free path becomes very long, in excess of 100 m, which typically exceeds the dimensions of the vacuum chamber by a large amount.

11.3.2 THE EQUIPARTITION OF ENERGY THEOREM

In the discussion leading up to Equation (11.4), it is established that the average translational kinetic energy of a molecule in an ideal gas is given by $\varepsilon_k = \dfrac{3}{2} k_B\,T$. In a 3D

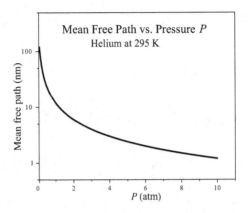

FIGURE 11.4 The pressure dependence of the mean free path l for helium at 295 K with pressures in the range 0 to 10 atm.

Cartesian frame of reference, there is an energy component associated with each of the three orthogonal directions. Since the axes are equivalent from symmetry considerations, it follows that $\varepsilon_{kx} = \varepsilon_{ky} = \varepsilon_{kz} = \frac{1}{2} k_B\, T$. The three energy contributions $\frac{1}{2} k_B\, T$ are associated with what are called the translational degrees of freedom. It is natural to ask whether polyatomic molecules have other internal energy contributions $\frac{1}{2} k_B\, T$ associated with their rotational and vibrational degrees of freedom. Experimental results and theoretical calculations have confirmed that this is the case provided the associated energy expression involves a *quadratic* dependence on an internal spatial coordinate. This, on average, equal sharing of energy among the various molecular degrees of freedom is known as the equipartition of energy theorem; it states that for a system in thermal equilibrium at temperature T, in the classical limit, each quadratic degree of freedom has a mean energy $\frac{1}{2} k_B\, T$.

As an example, consider a diatomic molecule such as oxygen. In addition to the three translational degrees of freedom, the molecule has two rotational degrees of freedom, and, in principle, two vibrational degrees of freedom corresponding to the potential and kinetic energies of vibration along the bond, giving a total of seven. However, the vibrational motion is generally non-classical at the temperatures of interest, and therefore the energy associated with this motion can be neglected. The two rotational degrees of freedom correspond to molecular rotations about orthogonal axes through the centre of mass, with energies of the form $\varepsilon_{rot} = \frac{1}{2} I\, \omega^2$. Rotational motion about the axis connecting the two atoms in a diatomic molecule does not contribute to the energy at the temperatures of interest, because of the extremely small moment of inertia about this axis. Figure 11.5 illustrates the two rotational degrees of freedom.

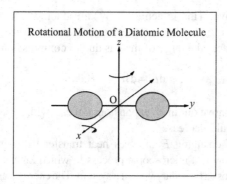

FIGURE 11.5 Rotational degrees of freedom for a diatomic molecule in a gas. Classical rotation occurs about the x and z axes as indicated.

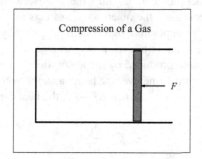

FIGURE 11.6 Compression of a gas using a piston-cylinder arrangement. Work is done on the gas as the piston of area A advances into the cylinder.

For an ideal gas of diatomic molecules, each with five degrees of freedom, the equipartition theorem predicts that, in equilibrium, the mean energy of a molecule is $\varepsilon = \dfrac{5}{2} k_\mathrm{B}\, T$.

11.4 THERMODYNAMIC PROCESSES: WORK AND HEAT

The internal energy E of a system can be changed in two distinct ways. Firstly, mechanical work can be done on the system by an external force. For example, the volume of a gas can be changed by moving a piston in a cylinder containing the gas, as depicted in Figure 11.6. This type of process is macroscopic in nature.

The mechanical work done by the force F in moving the piston through a small distance dx is $F\,dx$. As shown in Chapter 7, the gas which is being compressed exerts an opposing force $F' = P\,A$ on the piston. If the system is kept close to equilibrium during the compression, then F and F' are almost equal. The magnitude of the infinitesimal work done on the gas in moving the piston is $dW = F\,dx = P\,A\,dx = P\,dV$. It is convenient to take the work dW done in compressing the gas, by *decreasing* its volume, as positive. This is achieved by introducing a minus sign as follows: $dW = -P\,dV$. Applying the law of mechanical energy conservation, the accompanying change in internal energy of the gas due to compression is given by

$$dE = dW = -P\,dV \qquad (11.7)$$

If the piston were to move out in an expansion process, with $dW < 0$, then the internal energy of the gas would decrease.

A second way of changing E involves heat transfer between the system and its surroundings. Heat flow is a microscopic process in which kinetic energy at the atomic level is transferred across a boundary wall of a system. The energy transfer process involves collisions of gas molecules with a wall of a container as depicted in Figure 11.2. Atoms in the wall acquire energy from an external source and transmit it to the neighbouring atoms, and eventually to other gas molecules in the system of interest.

Thermodynamics does not enquire into the nanoscale details of heat transfer processes, but simply considers the macroscopic effects produced by heat flow, and specifically the change in temperature of a system produced by the absorption of heat. The heat capacity of a system is of central importance in this approach. The rise in temperature ΔT of a system produced by the absorption of heat ΔQ from a source is written as $\Delta Q = C\,\Delta T$, with C the average heat capacity over the temperature range involved. In the infinitesimal limit, when $\Delta T \to 0$, the heat capacity is defined as

$$C = \frac{dQ}{dT} \qquad (11.8)$$

For gases, in contrast to liquids and solids, the heat capacity of a gas depends on whether the measurements are made at constant volume or constant pressure. Further details are given below.

In discussing heat transfer, it is often convenient to introduce the concept of what are termed heat baths. A heat bath is a very large physical system that can give up or receive heat without a detectable change in its temperature. Heat baths have extremely large heat capacities. The change in the internal energy dE of a system due to heat dQ entering or leaving it, is given by the law of energy conservation as

$$dE = dQ \qquad (11.9)$$

Heat added to a system is taken to be positive, while heat extracted from a system is negative.

11.5 THE FIRST LAW OF THERMODYNAMICS

From Equations (11.7) and (11.9), the following relationship is obtained for the infinitesimal change in the internal energy of a system to which heat dQ is added and on which work $dW = -P\,dV$ is done,

$$dE = dQ + dW = dQ - P\,dV \qquad (11.10)$$

This is the mathematical statement of the first law of thermodynamics, which is an expression of the law of energy conservation. The first law can be adapted to apply to a wide variety of systems. As a starting point, the discussion below focuses on gases, and in particular the ideal gas.

11.5.1 THE FIRST LAW FOR AN IDEAL GAS

From the equipartition theorem, the internal energy of an ideal monatomic gas with three molecular degrees of freedom is given by Equation (11.4) as $E = \frac{3}{2}nRT$. This result is readily extended to diatomic molecules, with two additional rotational degrees of freedom. For a process carried out at constant volume, the first law becomes $dE = dQ$. The heat capacity at constant volume is obtained using Equation (11.8), with the addition of a subscript to indicate the constant volume constraint, as

$$C_V = \left(\frac{dQ}{dT}\right)_V = \left(\frac{dE}{dT}\right)_V = \frac{3}{2}nR \tag{11.11}$$

Replacing dE by $C_V \, dT$, the first law for an ideal gas becomes

$$C_V \, dT = dQ - P \, dV \tag{11.12}$$

Equation (11.12) has been obtained for a monatomic ideal gas, but it is readily adapted to a gas of polyatomic molecules with f degrees of freedom.

Exercise 11.3: Obtain an expression for the work done and the heat transferred to a heat bath at temperature T in the isothermal *compression* of an ideal gas from an initial volume V_i to a final volume V_f. Figure 11.7 depicts the work done on the gas and the heat transferred in the isothermal process.

For an isothermal process $C_V \, dT = 0$, and Equation (11.12) becomes $0 = dQ - P \, dV$. Making use of the ideal gas law $PV = nRT$ to replace P, and

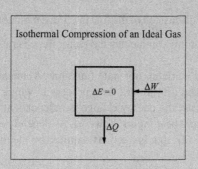

FIGURE 11.7 The work done on an ideal gas during isothermal compression is matched by the heat rejected to a heat bath (not shown).

then integrating over the volume change, gives the heat transferred to the heat

bath as $\Delta Q = n\,RT \int_{i}^{f} \dfrac{dV}{V} = n\,RT \ln\left(\dfrac{V_f}{V_i}\right)$. The first law then gives

$$\Delta W = -\Delta Q = -n\,RT \ln\left(\frac{V_f}{V_i}\right) \tag{11.13}$$

Since the volume has decreased, it follows that $\Delta W > 0$ while $\Delta Q < 0$, showing that heat is emitted to the heat bath at temperature T during the compression process in order to keep the gas's temperature constant.

11.5.2 GENERALIZED FORM OF THE FIRST LAW

For systems other than gases, it is necessary to express the work done on or by a system in terms of the mechanical variables which apply to the particular system. For example, consider a wire subject to a stretching force F which produces an increase in length dl. The work done in the process is $dW = F\,dl$. In general, the infinitesimal work done is written as $dW = Y\,dX$, with Y a generalized force and X a generalized displacement. The first law then has the form $dE = dQ + Y\,dX$. This modified first law is useful in a wide variety of applications from soap films to magnetic materials.

11.5.3 STATE VARIABLES AND STATE FUNCTIONS

In an ideal gas system, the variables which specify the state of the system are pressure P, volume V, and absolute temperature T. Since the state variables are connected by the ideal gas law, which is an equation of state, it follows that any two variables fix the value of the third variable, and thus are sufficient to specify the state of the system.

A state function can be expressed in terms of state variables. The internal energy of an ideal gas is a state function of the absolute temperature, as shown in Equation (11.4), and as used above in the discussion of the first law. For a monatomic gas $E(T) = \dfrac{3}{2} n\,RT$, while for a gas of polyatomic molecules the factor three is replaced by the number of degrees of freedom f.

In contrast, heat and work are not state functions. Mathematically, dE is an exact differential, while dQ and dW are not. In a process in which a system goes from an initial state to a final state, the energy change is independent of the path followed in the process. The work and heat inputs, however, do depend on the path followed, and it is only the sum $\Delta Q + \Delta W$ that is fixed, as required by the first law.

11.5.4 THE HEAT CAPACITY RELATIONSHIP

Equation (11.11) gives the heat capacity of a monatomic ideal gas as $C_V = \dfrac{3}{2} n\,R$. Using the first law, a relationship between C_V and the heat capacity at constant pressure C_p is established as follows. From Equation (11.12), the first law is written as $dQ = C_V\, dT + P\, dV$. The heat capacity at constant pressure is then given by $C_p = \left(\dfrac{dQ}{dT}\right)_P = C_V + P\left(\dfrac{\partial V}{\partial T}\right)_P$. Using the ideal gas law gives $P\left(\dfrac{\partial V}{\partial T}\right)_P = nR$. The difference of the heat capacities is thus given by

$$C_P - C_V = nR \qquad (11.14)$$

For a monatomic gas, $C_V = \dfrac{3}{2}nR$ and Equation (11.14) gives $C_p = \dfrac{5}{2}nR$. The physical reason for C_p being larger than C_V is that in a constant pressure ("isobaric") process, work is done to keep P constant as the system expands or contracts, while in a constant volume ("isochoric") process, no work is involved since V does not change.

The specific heat of a system c is defined as the heat capacity per unit quantity of the substance. The quantity is often taken to be per mole with $c = C/n$, or, alternatively, per unit mass with $c = C/m$. For gases, it is always necessary to distinguish between c_V and c_p. In contrast, this distinction is less important for liquids and solids, which are much less compressible than gases. For reference purposes, the specific heat per mole of a monatomic gas at constant volume is $c_V = \dfrac{C_V}{n} = \dfrac{3}{2}R = 12.47$ J mol^{-1} K^{-1}.

11.6   P–V DIAGRAMS

For gas systems, it is instructive to represent thermodynamic processes using P–V diagrams. As is noted above, just two state variables are sufficient to specify the state of a gas that obeys the ideal gas law. Figure 11.8 shows a representative P–V diagram for a gas undergoing an isobaric (constant pressure) process between initial and final volumes. Isochoric (constant volume) processes (not shown) would be represented by vertical lines, while the isobaric process is shown as the horizontal line. The work done on the gas in the isobaric process is $\Delta W = -\int_i^f P\, dV$, and, with attention to units, this is given by the area under the horizontal line. No work is done in isochoric processes, which involve temperature changes produced by heat flow.

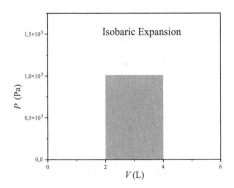

FIGURE 11.8 *P–V* diagram for a gas undergoing an isobaric expansion process. The work done in the process is given by the shaded area shown.

Exercise 11.4: The *P–V* diagram in Figure 11.8 represents the isobaric expansion of 0.1 mol of an ideal monatomic gas from an initial volume of 2 L to a final volume of 4 L with the pressure kept at 1 atm. Calculate the temperature change of the gas, the heat absorbed, and the work done *by* the system.

The temperature change is obtained using the ideal gas law as follows:

$$\Delta T = \left(\frac{P}{nR}\right)\Delta V = 1.219\times10^5 \times 2\times10^{-3} = 244 \text{ K}. \text{ The heat absorbed by the}$$

gas is $\Delta Q = C_p \, \Delta T = \frac{5}{2} n \, R \, \Delta T = 2.079\times244 = 507$ J. The work done by the gas is $\Delta W = P \, \Delta V = 101325\times2\times10^{-3} = 203$ J.

Note that the work done is given by the shaded area in Figure 11.8 expressed in SI units. The heat absorbed by the gas in the isobaric expansion process is largely converted into internal energy and the balance into work done by the system.

Exercise 11.5: Construct a *P–V* diagram to depict an isothermal compression process for an ideal gas. Take *T* = 295 K, *n* = 0.1 mol, and the initial and final volumes as 8 L and 1 L, respectively.

The *P–V* diagram for the isothermal process is readily obtained using the ideal gas law in the form $P = n \, R \, T/V = 2.45\times10^5/V$ Pa, with volumes given in L, as shown in Figure 11.9.

The work done on the gas is obtained using Equation (11.13), which gives $\Delta W = n \, R \, T \ln\left(\dfrac{V_i}{V_f}\right) = 0.1\times8.314\times295\times\ln(8) = 510$ J.

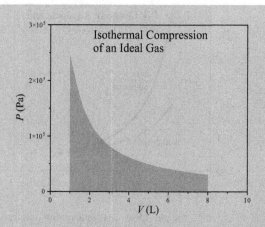

FIGURE 11.9 P–V diagram for the isothermal compression of 0.1 mol of an ideal gas from 8 L to 1 L at $T = 295$ K. The shaded area, with attention to units, is a measure of the work done on the gas.

11.7 ADIABATIC PROCESSES

In addition to isothermal, isobaric, and isochoric processes, there is a fourth important thermodynamic process known as the adiabatic process. In an adiabatic process, no heat is transferred to or from the system concerned, so that $\Delta Q = 0$. In contrast to what is known as Boyle's law, with $P\,V = $ constant for an ideal gas undergoing an *isothermal* process, it is found that for an *adiabatic* process, $PV^{\gamma} = $ constant with the exponent γ taking values determined by the number of degrees of freedom of the gas molecules. This form is discussed in detail below. In an adiabatic process carried out on an ideal gas, the state variables P, V, and T all change subject to the constraint that the ideal gas law connecting these variables always holds.

From Equation (11.12), the first law of thermodynamics can be written as $dQ = C_V\,dT + P\,dV$. Introducing the differential $d(PV) = P\,dV + V\,dP = nR\,dT$, based on the ideal gas law, gives $P\,dV = -V\,dP + nR\,dT$. Substituting in the first law equation leads to the relationship $dQ = C_V\,dT - V\,dP + nR\,dT = C_p\,dT - V\,dP$, where use has been made of the identity $C_p - C_V = nR$ given in Equation (11.14).

For an adiabatic process $dQ = 0$, and two alternative forms of the first law are, firstly,

$$P\,dV = -C_V\,dT \tag{11.15a}$$

and, secondly,

$$V\,dP = C_p\,dT \tag{11.15b}$$

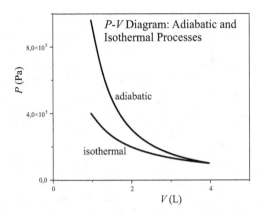

FIGURE 11.10 P–V diagram for an ideal gas undergoing adiabatic and isothermal compression processes from a volume of 4 L to 1 L. The internal energy of the gas increases in the adiabatic process, but not in the isothermal compression.

Dividing Equation (11.15b) by Equation (11.15a), and then rearranging, results in the differential equation $\dfrac{dP}{P} = -\dfrac{c_p}{c_V}\dfrac{dV}{V}$. Integration of this equation yields

$\ln P = -\dfrac{c_p}{c_V}\ln V + \ln\left(\text{constant}\right)$, and taking antilogs results in the form

$$P\,V^\gamma = \text{constant} \tag{11.16}$$

with $\gamma = c_p/c_V$. For a monatomic ideal gas $c_V = \dfrac{3}{2}R$ and $c_p = \dfrac{5}{2}R$, giving $\gamma = 5/3$. Using the ideal gas law, Equation (11.16) can also be written in the alternative form $T\,V^{\gamma-1} = \text{constant}$.

Figure 11.10 gives a P–V diagram for 0.16 moles of an ideal gas, showing both an adiabatic process with $PV^\gamma = \text{"constant 1"}$ and an isothermal process at 295 K with $PV = \text{"constant 2"}$. The constants were chosen to make the pressures equal for $V = 4$ L. Note that the pressure rises more rapidly with decreasing volume in the adiabatic process than in the isothermal case. Work done on the gas in the adiabatic compression increases the internal energy, as required by the first law for $\Delta Q = 0$. The temperature of the gas thus increases, since $E = \dfrac{3}{2}nRT$, and this enhances the pressure increase. No increase in temperature occurs in the isothermal process, which involves a heat bath to absorb heat from the system.

A combination of adiabatic and isothermal processes can be used to generate a heat–work cycle in which an ideal gas absorbs heat from a heat source, converts a portion into work output, and discards the remainder to a heat sink. A cycle of this type provides a model for a heat engine as discussed in Chapter 12.

11.8 THE SPECIFIC HEAT OF SOLIDS

The present chapter has introduced the first law of thermodynamics and has illustrated its use by considering processes involving an ideal gas. Expressions have been obtained for the specific heats at constant volume and constant pressure. In the nineteenth century, Dulong and Petit found that the molar-specific heats of most solids at standard temperature and pressure obeyed what has been called the Dulong–Petit law, in the form

$$c_p = 3R \tag{11.17}$$

For solids, there is in general little difference between c_p and c_V because of the small thermal expansion coefficients for these materials.

Equation (11.17) is strikingly similar to the molar-specific heat of a monatomic ideal gas, $c_V = \dfrac{3}{2}R$. The equipartition theorem provides an explanation for the high-temperature specific heats of solids using a model in which the atoms are connected to neighbours by springs. As the atoms in the solid vibrate about their equilibrium positions, they possess both kinetic and potential energy. The number of degrees of freedom is thus six, and not three as in a gas.

Further experimental measurements showed that the specific heats of solids decreased smoothly towards zero at low temperatures. It became clear that the law of Dulong and Petit no longer held at low temperatures, and the specific heat behaviour of a particular solid with temperature depends on its mechanical properties. In the early twentieth century, simple models, known as the Einstein and Debye models, together with quantum physics developments, provided an explanation for the observed low-temperature specific heat behaviour.

12 Entropy and the Second Law

12.1 INTRODUCTION

The internal energy of a system is of fundamental importance in considering its physical properties. As discussed in Chapter 11, the first law of thermodynamics, which is based on the law of energy conservation, relates the change in internal energy of a system to the heat and work it exchanges with its surroundings. From the heat capacity properties, it is shown that the internal energy E of an ideal gas is a state function of the state variable T. It is natural to ask if there are other state functions or state variables. In this chapter, it is shown that a quantity called the entropy is a state function of fundamental importance in thermal physics. The second law of thermodynamics relates the entropy changes of a system undergoing a thermodynamic process to the reversible or irreversible nature of the process. At a fundamental level, an increase in the entropy of a system is related to an increase in its disorder.

The entropy concept emerged over a period of many years following the theoretical analysis of heat engine operation in the early part of the nineteenth century. A deeper understanding of entropy came with the development of statistical mechanics later in the century. This chapter uses the operation of heat engines as the basis for introducing the entropy concept and the second law of thermodynamics.

12.2 HEAT ENGINES

Heat engines are mechanical devices, which, operating in a cycle, convert a fraction of the heat absorbed from a high-temperature source into useful work, and reject waste heat to a low-temperature sink. In analysing an engine, it is necessary to calculate the work done and the heat transferred per cycle. The following application illustrates the general approach.

DOI: 10.1201/9781003485537-12

Application 12.1: Consider a simple heat engine using a monatomic ideal gas as its working substance. The gas is contained in a cylinder, with a movable piston that can be clamped for isochoric processes and unclamped for isobaric processes. Heat baths, which can be placed in thermal contact with the cylinder, provide fixed temperatures at the four points shown in Figure 12.1. Find the work done per cycle, when operated in a clockwise sense, and the heat transferred in each process. Note the use of practical units ("atm L") below, rather than SI units (1 atm L = 101.325 J).

The work done on the ideal gas per cycle is given by $W = -\int\limits_{cycle} P\,dV =$

$-P_2\left(V_b - V_a\right) - P_1\left(V_d - V_c\right) = -2 \times 1 + 1 \times 1 = -1$ atm L. Converting the units using 1 atm L = 101 J gives $W = -101$ J. The sign convention used measures work *output* as negative. Note that the work output per cycle, in units "atm L", is given by the area of the shaded region in Figure 12.1.

The heat transfer in each process is given by $\Delta Q = C\,\Delta T$, with the heat capacity given by either $C_P = \dfrac{5}{2}R$ (isobaric) or $C_V = \dfrac{3}{2}R$ (isochoric) as derived for an ideal gas in Chapter 11. Heat is added to the system along paths a → b and d → a, while heat is removed along paths b → c and c → d. It is necessary to determine the temperature change along each path. This is done using the ideal gas law in the form $T = \dfrac{PV}{nR}$. The calculations are simplified by choosing $n = 0.120$ mol, which gives $nR = 1$ J/K. The following temperatures are obtained: $T_a = 606$ K, $T_b = 808$ K, $T_c = 404$ K, and $T_d = 303$ K. Heat added along isobaric path a → b is given by $Q_{ab} = \dfrac{5}{2}nR\,\Delta T_{ab} = \dfrac{5}{2} \times 202 = 505$ J.

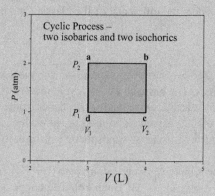

FIGURE 12.1 P–V diagram for a cyclic process consisting of two isobaric and two isochoric processes. Pressures are given in atm and volumes in L. The cycle operates in the clockwise sense, a → b → c → d.

Similar calculations give $Q_{bc} = -606$ J, $Q_{cd} = -252.5$ J, and $Q_{da} = 454.5$ J. Note that the *net* heat input per cycle is $Q_{cycle} = 101$ J, which is equal to the work done per cycle given above.

12.3   HEAT ENGINE EFFICIENCY

Consider a heat engine operating in a cycle, with heat input Q_1, work output W, and discarded waste heat Q_2 in each cycle. If a heat engine can be run backwards, it becomes a heat pump or refrigerator with the inputs and outputs switching signs. An important measure of the performance of a heat engine is its efficiency η, which is defined as

$$\eta = \frac{W}{Q_1} \tag{12.1}$$

The use of the first law of thermodynamics gives $\Delta E_{cycle} = \Delta Q - W$. Signs are chosen positive for inputs and negative for outputs, as used in Chapter 11. Since E is a state function, it follows that $\Delta E_{cycle} = 0$, because in a complete cycle the system returns to its initial state. Over a cycle, the first law expression gives $W = \Delta Q = Q_1 + Q_2$ with the signs of Q_1 and Q_2 determined by calculation as shown in Application 12.1. The heat engine efficiency, defined in Equation (12.1), becomes

$$\eta = \frac{Q_1 + Q_2}{Q_1} = 1 + \frac{Q_2}{Q_1} \tag{12.2}$$

Equation (12.2) has an important role in considering the operation of heat engines. Application 12.2 considers the efficiency of the classic Otto cycle as an illustrative example.

Application 12.2: The Otto cycle is a model for the operation of gasoline powered heat engines. The basic clockwise cycle involves four processes, two adiabatics (paths 1 → 2 and 3 → 4) and two isochorics (paths 2 → 3 and 4 → 1) as shown in Figure 12.2. Air intake at the start of each cycle and exhaust gas emission at the end both occur at atmospheric pressure. These two processes can be ignored.

In each cycle heat input Q_1 occurs along path 2 → 3 (V_2 isochore) followed by heat output Q_2 along 4 → 1 (V_1 isochore). The temperatures at points 1 to 4 in Figure 12.2 are designated T_1 to T_4, with $T_3 > T_2$ and $T_4 > T_1$. Obtain an

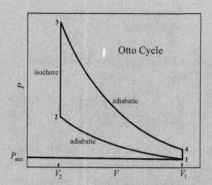

FIGURE 12.2 P–V diagram for the Otto heat engine cycle, which involves two adiabatics and two isochorics. Heat input Q_1 occurs along path $2 \rightarrow 3$ at volume V_2, while heat output Q_2 occurs along path $4 \rightarrow 1$ at volume V_1. Adiabatic compression happens along $1 \rightarrow 2$, while adiabatic expansion, producing work output, takes place along $3 \rightarrow 4$, which is the power stroke. Gas intake and exhaust processes take place at pressures close to atmospheric.

expression for the efficiency η of the cycle in terms of the volumes V_1 and V_2. Express the efficiency in terms of the compression ratio $R = V_1/V_2$.

Equation (12.2) gives the efficiency as $\eta = 1 + Q_2/Q_1$. The two heat exchanges occur along the isochoric paths $2 \rightarrow 3$ and $4 \rightarrow 1$. Taking initial and final temperatures around a cycle gives $Q_1 = C_V\left(T_3 - T_2\right) > 0$ and $Q_2 = C_V\left(T_1 - T_4\right) < 0$, where C_V is the heat capacity at constant volume for the gas used in the Otto cycle. Substituting the expressions for Q_1 and Q_2 into the equation for η gives,

$$\eta = 1 - \frac{T_4 - T_1}{T_3 - T_2} \tag{12.3}$$

Note the sign switches that have been made, to emphasize that $\eta \le 1$. Later in this chapter, it is shown that it is impossible for a heat engine to achieve $\eta = 1$.

Expressions for the positive temperature differences in Equation (12.3) are obtained by adapting the adiabatic relationship $P V^{\gamma} = \text{constant}$ given in Equation (11.16). Using the ideal gas law in the form $P = n RT/V$, the adiabatic relationship is rewritten as

$$T V^{\gamma-1} = \text{constant} \tag{12.4}$$

Applying Equation (12.4) to the two adiabatic processes gives $T_1 V_1^{\gamma-1} = T_2 V_2^{\gamma-1}$ along $1 \rightarrow 2$ and $T_3 V_2^{\gamma-1} = T_4 V_1^{\gamma-1}$ along $3 \rightarrow 4$. Subtracting the first of these equations from the second, and grouping terms in V_1 and V_2, leads to

$$\left(T_4 - T_1\right)V_1^{\gamma-1} = \left(T_3 - T_2\right)V_2^{\gamma-1} \qquad (12.5)$$

Combining Equations (12.3) and (12.5) gives the Otto cycle efficiency as

$$\eta = 1 - \left(\frac{V_2}{V_1}\right)^{\gamma-1} \qquad (12.6)$$

In terms of the compression ratio $R = V_1/V_2$, the efficiency is $\eta = 1 - 1/R^{\gamma-1}$. The higher the compression ratio, the higher the efficiency of the heat engine, which reaches 60% for a compression ratio $R = 10$, assuming $\gamma = c_p/c_V = 1.4$, which approximates the ratio of specific heats for air. In a practical gasoline engine, there are limitations on how high the compression ratio can be made without encountering problems due to preignition of the air–fuel mixture.

12.4   CARNOT CYCLE

12.4.1   CARNOT HEAT ENGINE

The Carnot cycle, which is named after Sadi Carnot who proposed it early in the nineteenth century, involves two adiabatic and two isothermal processes carried out *reversibly* on an ideal gas. All processes are quasi-static to ensure that the system is always close to equilibrium. The cycle depicted in the P–V diagram given in Figure 12.3 is reversible, and can operate either as a heat engine, taking in heat and performing work, or in reverse, as a refrigerator or a heat pump in an air conditioner.

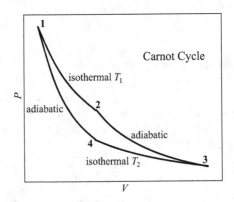

FIGURE 12.3   P–V diagram for the Carnot heat engine. Each cycle involves two isothermal and two adiabatic processes. Heat transfer takes place in the isothermal stages, and temperature changes occur along the adiabatics.

The efficiency of a Carnot heat engine is obtained using Equation (12.2) which involves the heat Q_1 absorbed in the *expansion* stroke 1 → 2, in which work is done by the system, and Q_2, the heat ejected in the *compression* 3 → 4 during which work is done on the system. As shown in Chapter 11, the heat Q_{if} transferred in the constant temperature expansion of an ideal gas from volume V_i to V_f is obtained using the first law with $\Delta E = 0$. This procedure gives

$$Q_{if} = W_{if} = \int_i^f P \, dV = n \, RT \ln\left(\frac{V_f}{V_i}\right) \tag{12.7}$$

Inserting the volume limits for processes 1 → 2 and 3 → 4 in a clockwise cycle, and then forming the ratio Q_2/Q_1, results in

$$\frac{Q_2}{Q_1} = \frac{T_2 \ln\left(V_4/V_3\right)}{T_1 \ln\left(V_2/V_1\right)} \tag{12.8}$$

It is straightforward to relate the volume ratios in Equation (12.7) to each other using the adiabatic relationship given in Equation (12.4). For the adiabatic process 2 → 3, $T_1 V_2^{\gamma-1} = T_2 V_3^{\gamma-1}$. Similarly for the adiabatic process 4 → 1, $T_2 V_4^{\gamma-1} = T_1 V_1^{\gamma-1}$, and dividing the first equation by the second, so that the temperatures cancel, gives $\left(V_2/V_1\right)^{\gamma-1} = \left(V_3/V_4\right)^{\gamma-1}$. Taking natural logarithms of both sides leads to $\ln\left(V_2/V_1\right) = -\ln\left(V_4/V_3\right)$, and comparison of this result with Equation (12.8) shows that $Q_2/Q_1 = -T_2/T_1$. Using Equation (12.2), the efficiency of a Carnot engine is obtained as

$$\eta = 1 - \frac{T_2}{T_1} \tag{12.9}$$

Note that the smaller the ratio T_2/T_1, the higher the efficiency. In the limit $T_2 \to 0$ K, the efficiency $\eta \to 1$. Achieving very low temperatures in a heat engine is impractical. In addition, experiment has shown that while very low temperatures can be attained using special equipment, it is impossible to reach 0 K as discussed below in Section 12.9. It is therefore impossible to achieve $\eta = 1$.

12.4.2 CARNOT HEAT PUMP

If a Carnot heat engine is run in reverse, it functions as a heat pump that extracts heat Q_2 from a low-temperature heat bath and rejects heat Q_1 at a high temperature. Work input W per cycle drives the heat pump. For a complete cycle, the first law, with $\Delta E = 0$, gives $W = -\left(Q_1 + Q_2\right)$. The coefficient of performance of the heat pump is defined as

$$\kappa = \frac{Q_2}{W} \tag{12.10}$$

The two heat transfers in a cycle are given by $Q_1 = n\,RT_1 \ln\left(V_1/V_2\right)$ and $Q_2 = n\,RT_2 \ln\left(V_3/V_4\right)$. Introducing the adiabatic equalities $T_2\,V_3^{\gamma-1} = T_1\,V_2^{\gamma-1}$ and $T_1\,V_1^{\gamma-1} = T_2\,V_4^{\gamma-1}$, and then dividing the first equation by the second with due regard for the signs of the logarithms gives $Q_2/Q_1 = -T_2/T_1$. The coefficient of performance becomes

$$\kappa = \frac{Q_2}{-Q_1 - Q_2} = \frac{T_2}{T_1 - T_2} \tag{12.11}$$

The value of κ exceeds unity and becomes very large when $T_1 - T_2 \ll T_2$. Reversible air-conditioners, which can operate as heat pumps, transfer heat from the cold outdoors to the warm interiors of buildings in winter. These devices provide a considerable saving in cost when compared with direct indoor electrical heating.

Application 12.3: Obtain an expression for the work done per cycle by a Carnot engine operating between heat baths at temperatures T_1 and T_2.

From the first law with $\Delta E = 0$ it follows that the work done per cycle is $W = Q_1 + Q_2$. Based on Equation (12.7), the heat *absorbed* in the isothermal expansion process $1 \rightarrow 2$ is given by $Q_1 = \int_1^2 P\,dV = n\,R\,T_1 \ln\left(V_2/V_1\right)$. The heat *rejected* in the isothermal compression $3 \rightarrow 4$ is obtained as $Q_2 = n\,R\,T_2 \ln\left(V_4/V_3\right)$. In addition, $\ln\left(V_2/V_1\right) = -\ln\left(V_4/V_3\right)$ follows by applying the adiabatic relationship given in Equation (12.4) to the two adiabatics, as shown above for the Carnot heat engine. Substituting the expressions for Q_1 and Q_2 into the equation for W, with due regard for the signs, gives

$$W = nR\left(T_1 - T_2\right)\ln\left(\frac{V_2}{V_1}\right) \tag{12.12}$$

Equation (12.12) shows that the work output per cycle depends on the temperature difference between the heat baths and on the natural logarithm of the volume expansion ratio V_2/V_1 in the isothermal expansion process.

12.5   ENTROPY AS A STATE FUNCTION

The energy of a thermodynamic system is a state function of the state variables as discussed in Chapter 11. For example, the energy of an ideal gas is a function of absolute temperature. The results obtained in the analysis of the Carnot cycle point the way to the introduction of another important state function called entropy. In a microscopic description, the entropy of a system is determined by the disorder in the

system. The greater the disorder, the higher the entropy. It follows that the entropy of
a material is higher in its gas phase than in its liquid or solid phases. Thermodynamics
provides a macroscopic description of processes involving work and heat. The entropy
changes that accompany thermodynamic processes are of primary interest in the present
discussion.

In determining the efficiency of a Carnot heat engine, the following result
is obtained: $\eta = 1 + Q_2/Q_1 = 1 - T_2/T_1$, with $Q_1 > 0$ and $Q_2 < 0$. It follows that
$Q_1/T_1 = -Q_2/T_2$, and in compact form this relationship becomes

$$\sum_{i=1}^{2} \frac{Q_i}{T_i} = 0 \qquad (12.13)$$

Equation (12.13) can be generalized to other closed cycle reversible heat engines
by using a set of Carnot cycles, which, when added together, produce a P–V diagram
which is a close approximation to that of the cycle considered. This procedure
assumes that hot and cold baths are available for the isothermal expansion and compression
processes in each Carnot cycle of the set. Adiabatic processes effectively
cancel one another in adjoining paths. Summing over the set of N Carnot cycles
involving $2N$ heat baths, leads to

$$\sum_{i=1}^{2N} \frac{Q_i}{T_i} = 0 \qquad (12.14)$$

Note that Equation (12.14) applies quite generally to *any* reversible cycle,
including a single Carnot cycle, in which heat transfer processes take place in
discrete small steps.

In order to simplify the notation, it is convenient to put $Q_i/T_i = \Delta S_i$ in Equation
(12.14) to give

$$\sum_{i=1}^{2N} \Delta S_i = 0 \qquad (12.15)$$

In the large N limit, the sum in Equation (12.15) can be converted to an integral over
a complete reversible cycle (denoted by the circle in the integral symbol below), with
the form

$$\oint dS = \int_{\text{cycle}} \frac{dQ}{T} = 0 \qquad (12.16)$$

Equation (12.16) is of fundamental importance in thermodynamics and, in honour
of the nineteenth-century physicist who established the result, it is known as
Clausius's theorem. As noted in Chapter 11, dQ is not an exact differential. The
relationship

$$\mathrm{d}S = \frac{\mathrm{d}Q}{T} \qquad (12.17)$$

gives $\mathrm{d}S$ as an exact differential, with $1/T$ the integrating factor for $\mathrm{d}Q$. The quantity denoted by the symbol S is termed the entropy.

The internal energy E of an ideal gas is a state function of the absolute temperature, as shown in Chapter 11. For a reversible Carnot cycle, $\oint \mathrm{d}E = 0$. The result $\oint \mathrm{d}S = 0$ given in Equation (12.16) for a reversible cycle provides compelling evidence that S is also a state function of the state variables.

The entropy concept that is introduced above is linked to the *change* in the state of a system. Equation (12.17) provides the basic relationship for calculating entropy changes accompanying a heat transfer process. Using the first law, the approach can be generalized to volume changes as described below. A general relationship for the entropy of a large system in terms of its huge number of accessible microstates Ω was obtained by Ludwig Boltzmann in the nineteenth century. The famous Boltzmann entropy equation is given by $S = k_{\mathrm{B}} \ln \Omega$. Using a microscopic approach to obtain an expression for $\Omega(E, V, N)$ leads to the Sackur–Tetrode equation for the entropy of an ideal gas of N molecules in a container of volume V. Further details are given in books on statistical physics. The present discussion deals with the entropy changes that occur in various thermodynamic processes.

12.6 ENTROPY CHANGES

Consider a system which makes a transition from an initial state i to a final state f with the states specified by state variables. While the process from i to f may be carried out in various ways, along paths which may be reversible or irreversible, the entropy change ΔS_{if} is independent of the path followed since entropy is a state function. It follows that an irreversible path can be replaced by a reversible path in calculating ΔS_{if}.

12.6.1 REVERSIBLE PROCESSES

In general, for a system with a temperature dependent heat capacity $C(T)$, the entropy change in a process from state i to state f is given by

$$\Delta S_{\mathrm{if}} = \int_{i}^{f} \frac{\mathrm{d}Q}{T} = \int_{i}^{f} \frac{C(T)\,\mathrm{d}T}{T} \qquad (12.18)$$

For many gases, and in particular the monatomic ideal gas, the heat capacity is temperature independent over a large temperature range as shown in Chapter 11. This feature simplifies the integral in Equation (12.18), giving the following result

$$\Delta S_{\mathrm{if}} = C \int_{i}^{f} \frac{\mathrm{d}T}{T} = C \ln\left(\frac{T_{\mathrm{f}}}{T_{\mathrm{i}}}\right) \qquad (12.19)$$

Equation (12.19) can be applied to *isochoric* and *isobaric* processes for a monatomic ideal gas, using the expressions for C from Chapter 11, $C_V = \frac{3}{2} n R$ and $C_P = \frac{5}{2} n R$ respectively.

Equation (12.19) does not apply to *isothermal* processes, in which a gas expands from volume V_i to V_f at constant T. In this case, Equation (12.7) gives $Q_{if} = n R T \ln\left(V_f / V_i\right)$, and hence

$$\Delta S_{if} = n R \ln\left(\frac{V_f}{V_i}\right) \qquad (12.20)$$

There is an accompanying change in the entropy of a heat bath, or reservoir, from which heat flows into (or out of) a gas system, as happens along the Carnot cycle path $1 \to 2$ shown in Figure 12.3. Because the heat capacity of the heat bath is taken to be extremely large, its temperature remains effectively constant at T. Since $Q_{bath} = -Q_{gas}$, the entropy change of the heat bath is equal in magnitude and opposite in sign to that of the gas.

Note that the entropy of a system depends on its size, as specified by its mass or by its number of moles of gas or some other material. Entropy is an *extensive* quantity with SI units J/K.

Application 12.4: Show that the total entropy change of a Carnot heat engine together with its heat baths is zero per cycle.

The total entropy change of the ideal gas and the two heat baths is given by $\Delta S = -\dfrac{Q_1}{T_1} + \dfrac{Q_1}{T_1} - \dfrac{Q_2}{T_2} + \dfrac{Q_2}{T_2} = 0$. The first term is the entropy loss $-Q_1/T_1$ of the hot bath, the second term the entropy gain Q_1/T_1 of the gas in the expansion stroke, the third term the entropy loss $-Q_2/T_2$ of the gas in the compression stroke, and the final term the entropy gain Q_2/T_2 of the cold bath. The four entropy changes sum to zero.

12.6.2 IRREVERSIBLE PROCESSES

An example of an irreversible process is the free expansion of an ideal gas from an initial volume V_i to a final volume V_f. A free expansion process can be carried out using the arrangement shown in Figure 12.4. No piston is involved, and the gas is initially in the container on the left, with the container on the right evacuated. Free expansion occurs when a seal covering an opening in the central partition is removed. No work is done in the expansion, and no heat is transferred. For an ideal gas, the temperature remains constant.

The entropy change of the gas in the isothermal expansion process is given by Equation (12.20) as $\Delta S_{if} = n R \ln\left(V_f / V_i\right)$. This follows because entropy is a state

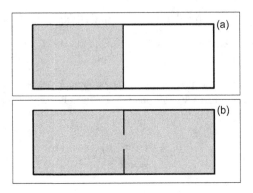

FIGURE 12.4 Free expansion of an ideal gas from an initial state (a) in which all the gas is in the container on the left while the container on the right is evacuated, to a final state (b) in which the gas fills both containers. Sudden expansion of the gas occurs after a seal in the central partition is removed. No work is done by the gas and no heat transfer occurs during the expansion process.

function, and the change in entropy depends only on the initial and final states of a system, and not on the path followed from state i to state f. Since $V_f > V_i$, it follows that $\Delta S > 0$.

To summarize, in a free expansion process, there is no work done and no heat is exchanged with a heat bath. The entropy increase of the gas is the only entropy change in the free expansion process. The entropy of the *universe* S_U has increased slightly by $\Delta S_U = \Delta S_{irr}$, because of the entropy increases ΔS_{irr} during the irreversible expansion process. This finding of an entropy increase, $\Delta S_U > 0$, in a free expansion process applies to all irreversible processes.

Application 12.5: A system initially consists of the ideal gases helium (He) and argon (Ar) in two containers separated by a partition, similar to the arrangement shown in Figure 12.4. The two containers are at the same temperature, with 0.2 moles of He in the left container and 0.2 moles of Ar in the right container. What is the change in entropy of the system produced by the removal of a seal covering an opening in the central partition, allowing the gases to mix? The pressures P of the gases in the two containers are equal before mixing takes place.

From the ideal gas law $PV = nRT$ it follows that the container volumes must be equal, since n, P, and T are initially the same for both He and Ar. Because the volumes of the two containers are equal, they are designated by V_i. After the mixing process each gas occupies the combined volume $V_f = 2V_i$. From Equation (12.20) the entropy of each gas increases by an amount $\Delta S_{if} = nR\ln\left(V_f/V_i\right) = nR\ln 2 = 1.15$ J/K.

The entropy change of the system is obtained by adding the entropy changes for each of the two gases, which together occupy both containers, leading to $\Delta S_{mix} = \Delta S_{He} + \Delta S_{Ar} = 2.3$ J/K.

Note that the mixed final state is disordered compared to the unmixed state. The loss of order produced by mixing is reflected in the entropy increase. The helium and argon atoms are intermingled and cannot be separated without considerable effort involving special equipment. The probability of ever finding the helium and argon atoms spontaneously restored to their original separate containers is vanishingly small.

Thermodynamic results similar to those obtained for the mixing of gases apply in the case of the free expansion of a gas. As shown above, the entropy increase is $\Delta S_{if} = n\,R\ln\left(V_f/V_i\right)$, consistent with a loss of order like the two-gases system. The probability of ever finding the gas in the state it was in prior to its free expansion is again vanishingly small.

Application 12.6: Two thermally insulated objects, 1 and 2, are initially at absolute temperatures T_1 and T_2, with $T_1 > T_2$. The objects are brought into thermal contact using a thermal link and allowed to reach equilibrium at final temperature T_f. Determine an expression for the entropy change in this irreversible process. Obtain the entropy change for $T_1 = 400$ K and $T_2 = 300$ K. Take the heat capacity of each bath as $C = 10^4$ J/K with negligible temperature dependence. Neglect heat losses to the surroundings.

The situation is depicted in Figure 12.5. In coming to equilibrium, the heat lost by object 1 is equal to the heat gained by object 2, giving $C\left(T_1 - T_f\right) = C\left(T_f - T_2\right)$. The equilibrium temperature is $T_f = \dfrac{1}{2}\left(T_1 + T_2\right)$.

The entropy change of object 1 is $\Delta S_1 = C\displaystyle\int_{T_1}^{T_f}\dfrac{dT}{T} = C\ln\left(T_f/T_1\right)$. Similarly,

for object 2 $\Delta S_2 = C\displaystyle\int_{T_2}^{T_f}\dfrac{dT}{T} = C\ln\left(T_f/T_2\right)$. The entropy change of the system

of two objects is $\Delta S = \Delta S_1 + \Delta S_2 = C\ln\left(T_f/T_1\right) + C\ln\left(T_f/T_2\right) = C\ln\left(\dfrac{T_f^2}{T_1\,T_2}\right)$.

For $\Delta S > 0$, the necessary condition is $\dfrac{T_f^2}{T_1\,T_2} = \dfrac{\left(T_1 + T_2\right)^2}{4T_1\,T_2} > 1$. This condition

reduces to $\left(T_1 - T_2\right)^2 > 0$, which is clearly satisfied.

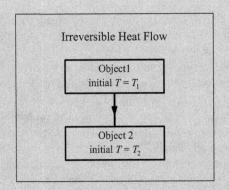

FIGURE 12.5 Irreversible heat flow occurs from hot object 1, initially at temperature T_1, to a cooler object 2, initially at temperature T_2. The entropy of the two-object system increases in this process. The heat capacities C of the two objects are assumed to be equal.

Using the initial temperatures of the objects, the entropy change which occurs in reaching thermal equilibrium is given by $\Delta S = 10^4 \times \ln\left(\dfrac{350^2}{300 \times 400}\right) = 206$ J/K.

12.6.3 Loss of Opportunity in Irreversible Processes

In addition to the increase in entropy which always accompanies an irreversible process, the opportunity to use the system to do useful work may be lost.

For example, in the situation involving two heat baths dealt with in Application 12.6, the opportunity to operate a heat engine between the two baths is lost once the two baths reach the final equilibrium temperature. While the two objects can be restored to their initial conditions, the increases in entropy cannot be reversed. The increase is just passed on to other systems involved in the restoration process.

12.6.4 The Local Universe

In considering the entropy changes that accompany a thermodynamic process, it is necessary to include changes that occur in the surroundings. For example, in dealing with a heat engine or heat pump, it is necessary to include heat baths. In order to ensure that all entropy changes are included, it is customary to specify the system together with its surroundings as the local universe. The total entropy change which occurs in a process is then called the entropy change of the local universe.

In processes such as photosynthesis, which involve energy reaching the Earth from the Sun, it is necessary, for completeness, to regard the Sun as part of the local universe. The nuclear fusion reactions in the solar interior are irreversible, and in five billion years, when its hydrogen fuel is exhausted, the Sun will become a white dwarf star.

12.7   THE SECOND LAW OF THERMODYNAMICS

Based on entropy considerations, the second law of thermodynamics is stated compactly as follows:

$$\Delta S_U \geq 0 \qquad (12.21)$$

This statement asserts that the change in entropy of the local universe associated with a system undergoing a thermodynamic process is either positive or zero. No decrease in entropy can occur in the local universe. Like other natural laws, this law is based on experiment and has never been found to be violated. The law is of fundamental importance in accounting for the behaviour of systems in many branches of science.

There are two historic statements of the second law, called the Clausius statement and the Kelvin–Planck statement, respectively. These alternative statements are based on heat engine considerations and are equivalent to the modern statement given above.

Since all irreversible processes lead to an entropy increase, it follows that the entropy of the universe is steadily increasing. The vast majority of natural processes are irreversible and are accompanied by growing disorder on length scales from the size of microscopic organisms to that of galaxies. While order can be restored locally, disorder continues its widespread increase. This endless increase in entropy is linked to the flow of time and provides what has been called time's arrow.

12.8   TEMPERATURE–ENTROPY DIAGRAMS

P–V diagrams, which were introduced in Chapter 11, provide instructive representations of thermodynamic processes in gases. The diagrams are used in this chapter to represent various cyclic processes, including the Carnot cycle. T–S diagrams provide an alternative graphical representation of thermal processes and are particularly well suited to showing adiabatics and isothermals.

The T–S diagram for a Carnot cycle given in Figure 12.6 has a rectangular shape made up of two isothermals and two adiabatics, with transitions between these processes occurring at the points labelled a to d. Note that in this representation, the heat engine cycle proceeds in an anticlockwise sense. The entropy scale is chosen to suit the situation that is being represented, with arbitrary zero entropy, since it is entropy *changes* that are of interest.

The efficiency of the Carnot engine is given by Equation (12.2) as $\eta = 1 + Q_2/Q_1$, where Q_1 is the heat absorbed at temperature T_1 along a $\rightarrow$ b, and Q_2 is the heat discarded at temperature T_2 along c $\rightarrow$ d.

For an isothermal process, from an initial state i to a final state f, integration of Equation (12.17) gives

$$\Delta Q_{if} = \int_i^f T \, dS = T \, \Delta S_{if} \qquad (12.22)$$

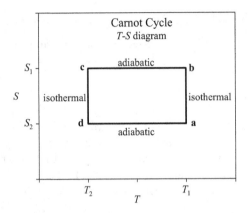

FIGURE 12.6   Carnot cycle T-S diagram showing two isothermal and two adiabatic processes. Heat Q_1 is absorbed along path a $\rightarrow$ b at temperature T_1, and heat Q_2 is discarded along path c $\rightarrow$ d at temperature T_2.

In the case of the Carnot cycle, Equation (12.22) gives $Q_1 = T_1 \Delta S_{ab} = T_1 \left(S_1 - S_2 \right)$ and $Q_2 = T_2 \Delta S_{cd} = T_2 \left(S_2 - S_1 \right)$. It follows that $\eta = 1 - T_2/T_1$ in agreement with Equation (12.9). Note that the heat input and heat output along the isothermals can be read off from the T–S diagram.

Application 12.7: Adapt the T–S diagram given in Figure 12.6 to the operation of a Carnot heat pump. Determine the coefficient of performance of the heat pump for $T_1 = 296$ K and $T_2 = 276$ K.

The T–S diagram for a Carnot heat pump has the same form as that shown in Figure 12.6, just with all the processes run in reverse. With the use of Equations (12.10) and (12.11), the coefficient of performance is given as $\kappa = \dfrac{Q_2}{W} = \dfrac{Q_2}{-Q_1 - Q_2} = \dfrac{T_2}{T_1 - T_2}$. Inserting the temperature values gives $\kappa = \dfrac{276}{20} = 13.8$. For every unit of work input to the heat pump, 13.8 units of heat are extracted from the cold bath.

12.9   THE THERMODYNAMIC LAWS

In addition to the first and second laws of thermodynamics, which have been introduced earlier, there are two other laws that play a role. The first of these is the zeroth law, also called the law of thermometry.

Law 0: If two separated systems are in thermal equilibrium with a third system, then they are in thermal equilibrium with each other.

As an illustrative example, consider a liquid-in-glass thermometer. If the liquid (e.g., mercury or alcohol) is in thermal equilibrium with the inner surface of the thermometer bulb, and the outer wall of the bulb is in thermal equilibrium with an object whose temperature is being measured, then the object and the liquid are in thermal equilibrium. The glass wall acts as an intermediate object, in thermal equilibrium with both the thermometer liquid and the system whose temperature is being measured. The thermometer is calibrated at fixed reference temperature points.

The third law of thermodynamics is concerned with the unattainability of absolute zero temperature, which is linked to the behaviour of a system's entropy at very low temperatures.

Law 3: The entropy of a system tends to zero as the temperature approaches zero kelvin.

Decreasing the entropy of a system corresponds to increasing its order. As the temperature of a system is lowered, the entropy continuously decreases, and squeezing the remaining entropy from the system becomes increasingly difficult. The third law implies that it is impossible to reach absolute zero in a finite number of steps using a succession of entropy-reducing processes. Ingenious experimental techniques have been developed in order to reach very low temperatures, well below 1 K. In a classic experiment, a gas of alkali atoms trapped by laser beams has been cooled to below 10^{-9} K.

The third law must be modified slightly when applied to inhomogeneous systems with intrinsic disorder. The generalized third law states that the entropy of a system tends to zero, or to a constant value, as the temperature approaches zero.

Compact statements of the four laws governing thermodynamic processes are given below.

Law 0: For 3 thermally interacting systems in equilibrium, if $T_1 = T_3$ and $T_2 = T_3$, then $T_1 = T_2$.

Law 1: The energy E of the local universe (LU) is constant, $\Delta E_{LU} = 0$.

Law 2: The entropy S of the LU is either constant or increasing, $\Delta S_{LU} \geq 0$.

Law 3: The entropy of a system tends to 0 as the temperature tends to 0 K, $S \rightarrow 0$ as $T \rightarrow 0\,\mathrm{K}$.

Note that a decrease in entropy governed by "law 3" is not inconsistent with "law 2". That is because the decrease in entropy of a particular system, which relies on the use of special techniques, is more than offset by an increase in entropy of the associated laboratory equipment, which is referred to as the local universe.

12.10 CONCLUSION

With the development of the laws of thermodynamics, this book has covered many of the physics topics of interest up until the mid-nineteenth century. At that time, experimental work was clarifying the phenomena of electricity and magnetism. This work culminated in Maxwell's equations of electromagnetism, which were developed in a set of papers published during the 1860s. A key implication of this work was that the speed of light should be the same in all inertial reference frames, which appears to be incompatible with Newton's conceptual framework that singles out an absolute reference frame determined by the fixed stars. Einstein resolved the incompatibility when he published the special theory of relativity for reference frames in uniform motion, a theory that was subsequently extended in the general theory of relativity to encompass reference frames undergoing acceleration. Einstein's theories generalize Newton's ideas to handle situations where a particle is travelling close to the speed of light or is in the vicinity of a very massive object.

At approximately the same time, another experimental issue, that of black body radiation, became known. Black body radiation is the electromagnetic radiation that is emitted by an opaque, non-reflective material which is in thermodynamic equilibrium with its surroundings. Classical physics could not explain the spectrum of the emitted radiation. The resolution of this issue eventually led to the development of quantum mechanics.

The theories of special and general relativity, and quantum mechanics, became the focus of much of physics research after the nineteenth century.

Appendices

APPENDIX 1: FUNDAMENTAL PHYSICAL CONSTANTS

Constant	Symbol	Value
Avogadro constant	N_A	$6.02214076 \times 10^{23}$ mol^{-1}
Boltzmann constant	k_B	1.380649×10^{-23} J $\cdot$ K^{-1}
Electron charge	e	$1.602176634 \times 10^{-19}$ C
Electron mass	m_e	$9.1093837139(28) \times 10^{-31}$ kg
Molar gas constant	R	8.31446261815324 J mol^{-1} K^{-1}
Gravitational constant	G	$6.67430(15) \times 10^{-11}$ m^3 kg^{-1} s^{-2}
Planck constant	h	$6.62607015 \times 10^{-34}$ J s
Proton mass	m_p	$1.67262192595(52) \times 10^{-27}$ kg
Speed of light in vacuum	c	299792458 m s^{-1}

APPENDIX 2: PERIODIC TABLE OF THE ELEMENTS

1																	18
1 1.008* **H** hydrogen	2											13	14	15	16	17	2 4.003 **He** helium
3 6.94* **Li** lithium	4 9.012 **Be** beryllium											5 10.81* **B** boron	6 12.01* **C** carbon	7 14.01* **N** nitrogen	8 16.00* **O** oxygen	9 19.00 **F** fluorine	10 20.18 **Ne** neon
11 22.99 **Na** sodium	12 24.31* **Mg** magnesium	3	4	5	6	7	8	9	10	11	12	13 26.98 **Al** aluminium	14 28.09* **Si** silicon	15 30.97 **P** phosphorus	16 32.06* **S** sulfur	17 35.45* **Cl** chlorine	18 39.95 **Ar** argon
19 39.10 **K** potassium	20 40.08 **Ca** calcium	21 44.96 **Sc** scandium	22 47.87 **Ti** titanium	23 50.94 **V** vanadium	24 52.00 **Cr** chromium	25 54.94 **Mn** manganese	26 55.85 **Fe** iron	27 58.93 **Co** cobalt	28 58.69 **Ni** nickel	29 63.55 **Cu** copper	30 65.38* **Zn** zinc	31 69.72 **Ga** gallium	32 72.63 **Ge** germanium	33 74.92 **As** arsenic	34 78.97* **Se** selenium	35 79.90* **Br** bromine	36 83.80 **Kr** krypton
37 85.47 **Rb** rubidium	38 87.62 **Sr** strontium	39 88.91 **Y** yttrium	40 91.22 **Zr** zirconium	41 92.91 **Nb** niobium	42 95.95* **Mo** molybdenum	43 [98] **Tc** technetium	44 101.1 **Ru** ruthenium	45 102.9 **Rh** rhodium	46 106.4 **Pd** palladium	47 107.9 **Ag** silver	48 112.4 **Cd** cadmium	49 114.8 **In** indium	50 118.7 **Sn** tin	51 121.8 **Sb** antimony	52 127.6 **Te** tellurium	53 126.9 **I** iodine	54 131.3 **Xe** xenon
55 132.9 **Cs** caesium	56 137.3 **Ba** barium	57–71	72 178.5 **Hf** hafnium	73 180.9 **Ta** tantalum	74 183.8 **W** tungsten	75 186.2 **Re** rhenium	76 190.2 **Os** osmium	77 192.2 **Ir** iridium	78 195.1 **Pt** platinum	79 197.0 **Au** gold	80 200.6 **Hg** mercury	81 204.4* **Tl** thallium	82 207.2 **Pb** lead	83 209.0 **Bi** bismuth	84 [209] **Po** polonium	85 [210] **At** astatine	86 [222] **Rn** radon
87 [223] **Fr** francium	88 [226] **Ra** radium	89–103	104 [267] **Rf** rutherfordium	105 [268] **Db** dubnium	106 [269] **Sg** seaborgium	107 [270] **Bh** bohrium	108 [277] **Hs** hassium	109 [278] **Mt** meitnerium	110 [281] **Ds** darmstadtium	111 [282] **Rg** roentgenium	112 [285] **Cn** copernicium	113 [286] **Nh** nihonium	114 [289] **Fl** flerovium	115 [290] **Mc** moscovium	116 [293] **Lv** livermorium	117 [294] **Ts** tennessine	118 [294] **Og** oganesson

*H: (1.00784, 1.00811)
Li: (6.938, 6.997)
C: (12.0096, 12.0116)
N: (14.00643, 14.00728)
O: (15.99903, 15.99977)
Mg: (24.304, 24.307)
Si: (26.084, 26.086)
S: (32.059, 32.076)
Cl: (35.446, 35.457)
Br: (79.901, 79.907)
Tl: (204.382, 204.385)
Zn: 65.38(2)
Se: 78.96(3)
Mo: 95.96(2)

57 138.9 **La** lanthanum	58 140.1 **Ce** cerium	59 140.9 **Pr** praseodymium	60 144.2 **Nd** neodymium	61 [145] **Pm** promethium	62 150.4 **Sm** samarium	63 152.0 **Eu** europium	64 157.3 **Gd** gadolinium	65 158.9 **Tb** terbium	66 162.5 **Dy** dysprosium	67 164.9 **Ho** holmium	68 167.3 **Er** erbium	69 168.9 **Tm** thulium	70 173.0 **Yb** ytterbium	71 175.0 **Lu** lutetium
89 [227] **Ac** actinium	90 232.0 **Th** thorium	91 231.0 **Pa** protactinium	92 238.0 **U** uranium	93 [237] **Np** neptunium	94 [244] **Pu** plutonium	95 [243] **Am** americium	96 [247] **Cm** curium	97 [247] **Bk** berkelium	98 [251] **Cf** californium	99 [252] **Es** einsteinium	100 [257] **Fm** fermium	101 [258] **Md** mendelevium	102 [259] **No** nobelium	103 [266] **Lr** lawrencium

Index

Printed in the United States
by Baker & Taylor Publisher Services